JANUA LINGUARUM

STUDIA MEMORIAE

NICOLAI VAN WIJK DEDICATA

edenda curat

C. H. VAN SCHOONEVELD

Indiana University

Series Minor, 197

FUNCTOR ANALYSIS OF NATURAL LANGUAGE

by

JOHN LEHRBERGER

1974

MOUTON

THE HAGUE · PARIS

LIBRARY OF CONGRESS CATALOG CARD NUMBER: 74-82387

Printed in Belgium, by N.I.C.I., Ghent

PREFACE

The central idea in this study is the analysis of the structure of a text in a natural language directly in terms of the relations between the phrases which make up that text. The vehicle for direct representation of these relations is the interpretation of some phrases in a text as functors and others as arguments of the functors. The analysis here is limited to English, but the basic methods should apply to other languages even though the details may differ.

In assigning functors within a text it is often helpful to compare that text with others which are paraphrases of it. Relations between texts are taken into account in chapter 7. Although only paraphrase is considered here, this does not mean that non-paraphrastic relations between texts are regarded as unimportant or unnecessary.

A grammar based on the kind of analysis proposed in this paper would be an extension of traditional categorial grammars. One important extension is the use of a single occurrence of a phrase as an argument of more than one functor in a text. Another extension is the assignment of more than one set of arguments to a single occurrence of a functor. These two extensions are referred to as argument sharing and functor sharing respectively and they are related to 'zeroing' in Harris's theory of transformations.

The first two chapters outline the historical development of categorial grammars and the use of the functor concept in the analysis of natural languages. It is pointed out that several early investigators recognized that certain category assignments were, in a sense, equivalent to the statement of transformational relations. The claim that categorial grammars are variants of phrase structure grammars is discussed in chapter two and the equivalence of several forms of categorial grammars is reviewed.

The topic of chapter three is the assignment of structure to a string of symbols and the representation of such structures by means of graphs. These are not derivation graphs. As a matter of fact, texts (or sentences) are not derived here either from kernel sentences or abstract structures. A method is given to generate graphs for structures in certain artificial languages. Another method is described for constructing graphs directly from texts to which functors and arguments have been assigned. The latter method is used to construct graphs for structured texts in natural language throughout the rest of the book.

In chapter four there is a discussion of criteria that might be useful in deciding which phrases in a given text should be regarded as functors initially. Suggestions are made, but no formal procedure is given. The idea is stressed that for a single reading of a text there is more than one possible assignment of functors and corresponding groupings of phrases in the text. This is related to the use of parametric forms of functors as presented in chapter three. The 'structure' corresponding to a given reading is therefore not a single functor-argument assignment, but a composite of all the possible assignments reflecting various 'interpretations' within that reading. These interpretations are related to focusing or topicalization.

Argument sharing and functor sharing are presented in chapter five. Examples are given involving referentials (which are treated here as functors) and then sharing is used to describe various cases of 'zeroing'. The graph representation becomes more complicated and 'trees' no longer suffice. Argument sharing leads to the introduction of cycles in the graphs and functor sharing requires the use of s-graphs.

In chapter six suprasegmental morphemes are treated as functors. The inclusion of such functors in the analysis emphasizes the non-stringlike nature of texts in a natural language and lends support to the use of graphs rather than strings as a notational device.

The relation between emphatic and contrastive stress is discussed and also the role of intonation as a functor. This chapter is not intended as an analysis of English stress and intonation; its main purpose is rather to point out that suprasegmentals need not and should not be neglected in a functor analysis of a natural language.

The structure of a language involves relations between texts as well as relations between phrases within a text. In chapter seven relations between structured texts are stated in terms of the functors which they contain. The term *transformation* is used, although not in the Harisian or Chomskian sense; it is closer to the usage of Zellig Harris. Not only is the definition different, but so is the role of transformations in the grammar. A transformation here is not a step in the derivation of a sentence or text; it is simply the recognition of a structural relation between certain paraphrastically related texts. Many sentences which are transformationally related by the criteria of chapter seven are also transformationally related by other definitions; some are not.

Terms from graph theory are defined in the appendix for convenience of reference.

CONTENTS

1

SEMANTIC CATEGORIES AND SYNTACTIC CONNEXION

1.1. A major concern of both linguists and logicians is the manner in which phrases in a language are combined to form other phrases in the language – the study of syntactic connection. The classification of phrases into grammatical categories is basic to this study. Of the various criteria that may be used to establish such categories the one which forms the historical starting point for the present study is the criterion of MUTUAL SUBSTITUTIVITY. Edmund Husserl proposed that the words and complex expressions of a language be classified on this basis and he called the resulting classes SEMANTIC CATEGORIES.[1] Roughly, the principle is that two expressions belong to the same semantic category if and only if each can be substituted for the other in a meaningful sentence and the result is a meaningful sentence. Thus in the sentence:

(1) Herman writes poems

we may replace *writes* with *reads* and the result is a meaningful sentence:

(2) Herman reads poems

It follows that *reads* and *writes* belong to the same semantic category by Husserl's definition. Note that it is MEANINGFULNESS, not meaning, that is preserved in the replacement.

In like manner *poorly* and *rapidly* would be placed in the same category since:

(3) Herman reads poorly

[1] Husserl, *Logische Untersuchungen*, vol. II, part I, *319*, *1.7*b ff.

and

(4) Herman reads rapidly

are both meaningful sentences. But we also have:

(5) Herman reads books

and

(6) Herman reads while the other boys play

This would mean, according to Husserl's definition, that *books* and *while the other boys play* belong to the same category as *poorly* and *rapidly*. Such a categorization would be unacceptable on both semantic and syntactic grounds.

Next consider the effect of changing Husserl's principle so that two expressions belong to the same semantic category if and only if the replacement of either expression by the other in EVERY meaningful sentence results in a meaningful sentence. That is, the replacement must be performable everywhere the expressions occur. Presumably, we would like for *reads* and *writes* to be in the same category. If they are, then the above principle tells us that since:

(7) Mary writes left-handed

is a meaningful sentence, so must be:

(8) Mary reads left-handed

Also, since we have:

(9) He reads in a loud voice

we must also have:

(10) He writes in a loud voice.

It seems that for almost any two expressions there is some sentence in which one expression is meaningful and the other is meaningless or at least questionable. If this is indeed the case, then the demand that replacement be everywhere performable is too strong.

From one point of view (8) and (10) are not meaningless, but strange. We might prefer to think in terms of ACCEPTABILITY rather than meaningfulness. Harris's transformations, e.g., are based on an acceptability grading among sets of sentences – relative, not absolute – which is preserved in the transformation.[2] Relative acceptability of two sentences is easier to decide than relative meaningfulness. We may say that (8) and (10) are less acceptable than (7) and (9) respectively, or perhaps acceptable in a different type of discourse. If we try to partition a set of sentences into meaningful and non-meaningful subsets we are surely going to run into difficulties. Of course, even if acceptability or relative acceptability is used as a criterion the examples of the preceding paragraph show that the resulting categories would be of little or no value.

Neither the demand that mutual replacement be performable in EVERY sentence nor in ONE sentence results in a useful classification. We may ask if there is some middle ground – a sufficiently large number of replacements without being exhaustive. Henry Hiż made a proposal to this effect in a paper *Congrammaticality, Batteries of Transformations and Grammatical Categories*.[3] He avoids the problem of 'every' by introducing the concept of the UNSPECIFIC LARGE QUANTIFIER THERE ARE MANY:

> To say, e.g. that in a large segmented text there are many segments, or discontinuous combinations of segments, satisfying a given condition means that a suitably large proportion of segments of the text, or of such combinations, satisfy the condition. ... The claim that there are many adjectives in Latin does not assert that they are infinitely numerous, but that the set of Latin adjectives constitutes, say, four percent of the total Latin vocabulary.

The linguistic segments are classified according to the way they enter batteries of transformations, and this yields grammatical

[2] Harris, *Mathematical Structures of Language*, 51-59.
[3] Hiż, "Congrammaticality, Batteries of Transformations and Grammatical Categories", in *Structure of Language and Its Mathematical Aspects*, ed. by Roman Jakobson, 43-44.

categories. The transformations are not set up initially in terms of strings of category symbols, but by direct comparison of actual sentences. (Sentences, taken as empirically given, are the starting point in this approach.) The replacement by segments (there are many and various replacements in each position) must preserve sentencehood. As more batteries of transformations are taken into account a more refined segmentation may occur and more refined categories result. Grammatical categories are then relative to context.

Whereas Husserl's doctrine is based on preservation of meaningfulness, Hiż bases his on the preservation of sentencehood. Neither doctrine depends on preservation of meaning, truth or information, hence they are both a-semantic.

1.2. Husserl's doctrine is also an ingredient in the theory of semantic categories advanced by the logician Stanislaw Leśniewski.[4] Leśniewski's theory was further elaborated by another logician, Kazimierz Ajdukiewicz, with changes in the symbolism. Our discussion will be based on the presentation in Ajdukiewicz's paper *Syntactic Connexion*.[5] He begins with the following precise definition of semantic category:[6]

> The word or expression A, taken in sense x, and the word or expression B, taken in sense y, belong to the same semantic category if and only if there is a sentence (or sentential function) S_A, in which A occurs with meaning x, and which has the property that if S_A is transformed into S_B upon replacing A by B (with meaning y), then S_B is also a sentence (or sentential function).

Since Ajdukiewicz's definition mirrors Husserl's it has the same flaws. This concept of semantic category is sentence-preserving, but not meaning-preserving. Note that if A and B are taken to be any two sentences, then we may take S_A as A and S_B as B and the definition is satisfied. Therefore all sentences belong to the same

[4] Leśniewski, "Grundzüge eines neuen Systems der Grundlagen der Mathematik".
[5] Ajdukiewicz, "Syntactic Connexion" (in *Polish Logic 1920-1939*), 207-231.
[6] Ajdukiewicz, "Syntactic Connexion", 208.

semantic category. The sentence category is one of the 'basic' categories.

A BASIC CATEGORY is any semantic category which is not a functor category. A FUNCTOR CATEGORY is any semantic category consisting of functors. A FUNCTOR[7] is described as an "unsaturated symbol" with "brackets following it". A functor takes one or more arguments of various categories and, together with its arguments, forms a phrase of some category. For example, in arithmetic '+' is a functor which forms a numerical phrase out of two arguments which are numerical phrases. The sentence category is not a functor category. Leśniewski and Ajdukiewicz both use two basic categories, SENTENCE and NAME. Ajdukiewicz notes that in natural languages names seem to fall into two categories, names of individuals and names of universals. There is nothing which prohibits the use of other basic categories.

The arguments of a functor are ordered. Functors are arranged in a hierarchy according to the number of arguments (taken in order), and then by the category of the resulting expression formed by the functor with its arguments.

To begin with, single words are assigned to either basic or functor categories. The index s is used for the sentence category, n for the name category and 'fractional' indices for functor categories. For example, the fractional index $\frac{s}{n}$ indicates a functor which forms an expression of the sentence category out of a single argument of the name category. The index $\frac{s}{ss}$ indicates a functor which forms a sentence out of two arguments which are both sentences. The proposition $\sim$ p would be analyzed as:

(11)
$$\begin{matrix} \sim & p \\ \frac{s}{s} & s \end{matrix}$$

and the implication $\supset$ pq would be analyzed as:

[7] Ajdukiewicz, "Syntactic Connexion", 209.

(12)
$$\begin{array}{ccc} \supset & p & q \\ \frac{s}{ss} & s & s \end{array}$$

There are many problems in the application of this method to natural languages. Nevertheless, Ajdukiewicz shows how the method might work in the following example:

(13) The lilac smells very strongly and the rose blooms:
$$\begin{array}{ccccccccc} \text{The} & \text{lilac} & \text{smells} & \text{very} & \text{strongly} & \text{and} & \text{the} & \text{rose} & \text{blooms} \\ \frac{n}{n} & n & \frac{s}{n} & \frac{\frac{\frac{s}{n}}{\frac{s}{n}}}{\frac{\frac{s}{n}}{\frac{s}{n}}} & \frac{\frac{s}{n}}{\frac{s}{n}} & \frac{s}{ss} & \frac{n}{n} & n & \frac{s}{n} \end{array}$$

lilac and *rose* belong to the basic category of names; *the* is regarded as a functor which forms a phrase of category n out of an argument of category n (hence *the lilac* and *the rose* are each of category n); *very* is a functor which takes *strongly* as its argument and forms a phrase of the same category as *strongly*; *very strongly* takes *smells* as its argument to form *smells very strongly*, of the same category as *smells*; and *the lilac* serves as argument of *smells very strongly*.

At this point the reader may wonder why *lilac* is not taken as argument of *smells*, forming a sentence *lilac smells*. How do we know what the arguments of a functor are in a given string? The letters in the 'denominator' of a fractional index are ordered like the arguments of the functor in the string, but this does not prevent taking *the lilac* as argument of *smells* rather than of *smells very strongly*. Furthermore, how do we know whether an argument is on the left or right of the functor? This presents no problem if, following Leśniewski, we always write the functor to the immediate left of its arguments. But natural languages are not so arranged. In (13) the functor *and* has one argument on its right *(the rose blooms)* and one on its left *(the lilac smells very strongly)*.

In order to analyze a sentence the SCOPE of each functor in the sentence must be known. Ajdukiewicz points out that in natural languages word order, inflections, prepositions and punctuation marks (intonation) all help in this respect. Leśniewski introduced the concept of the MAIN FUNCTOR of an expression: when an expression is divided into parts such that one part is a functor and the remaining parts are its arguments, that functor is called the main functor of the expression. In (11) *and* is the main functor. An expression which can be divided into a main functor and its arguments is said to be WELL ARTICULATED.

Suppose an expression E is well articulated. The main functor of E and each of its arguments may also be well articulated expressions. If we continue this process, looking at the parts of the parts, the parts of these, etc., and at each stage every part is well articulated, until finally we reach parts which are all single words, then E is said to be WELL ARTICULATED THROUGHOUT.

This kind of segmenting of an expression is similar to the segmentation in IMMEDIATE CONSTITUENT ANALYSIS. We divide an expression into parts, sub-divide the parts, etc. There is a hierarchical construction in both cases. In IC analysis the first segmentation of an expression gives its immediate constituents just as the first segmentation in a functor analysis gives the main functor and its arguments. Ajdukiewicz refers to these segments as FIRST ORDER PARTS. However, the main functor and its arguments need not be the same as the immediate constituents of an expression, nor is the number of segments necessarily the same by each method.

Ajdukiewicz calls the main functor and its arguments FIRST ORDER PARTS of the expression. Just as in IC analysis, we may also speak of Kth ORDER PARTS. Thus in (13) *and* is a first order part, *the rose blooms* is a first order part, *the rose* and *blooms* are second order parts, *the* and *rose* are third order parts.

The analogy between IC analysis and functor analysis must not be pushed too far. In addition to the fact that the number and kind of segments may not be the same, there is an even more important way in which the two methods differ. Functor analysis is more than just a segmentation. At each stage the segments consist of a main functor and its arguments. If such a segmentation is represented by a familiar tree diagram (like the ones used in studying phrase structure grammars), the branches from a node will not all be the same. The branch leading to the main functor must be distinguished from the branches leading from that same node to the arguments of the main functor.[8] For example, if an expression E consists of main functor F and arguments A_1, A_2, then the tree diagram is not (14), but (15).

(14)

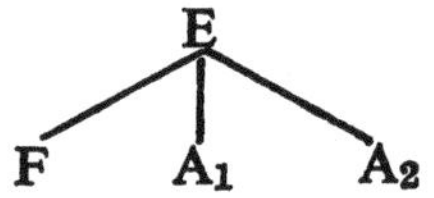

(15)

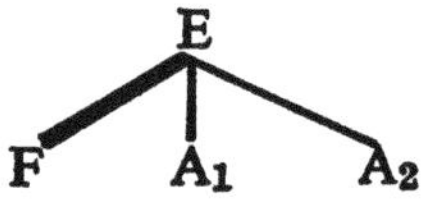

The segmentation of an expression into main functor and arguments establishes a certain relation between the segments. The argument segments have to meet the conditions specified by the fractional index of the functor segment. Not only must there be the exact number of kth order parts to serve as arguments for each of the kth order main functors (no segment serving as an argument for more than one functor), but they must belong to the proper categories specified in the denominator of the fractional index of the corresponding functor: If a main functor F which is a kth order part of E has an index:

$$\frac{C_0}{C_1 \ldots C_n}$$

then there must be n other k^{th} order parts $A_1, \ldots, A_n$ such that A_1 is the first argument of F and belongs to the category $C_1, \ldots, A_n$ is the nth argument of F and is of category C_n; and none of the segments $A_1, \ldots, A_n$ is an argument of any functor other than F.

If an expression is well articulated throughout and all the preceding conditions are met, then that expression is said to be

[8] This was pointed out by H. Hiż.

SYNTACTICALLY CONNECTED. (Syntactic connection will be discussed further in 6.3 following developments in chapters 3-6.)

Ajdukiewicz anticipates the necessity of considering some sentences in a natural language as *transforms* of others via 'zeroing'[9] (although he does not use those terms):[10]

> ... ordinary language often admits elliptical expressions so that sometimes a significant composite expression cannot be well articulated throughout on the sole basis of the words explicitly contained within it. But a good overall articulation can be easily established by introducing the words omitted but implicit.

He also mentions the problem of 'separable words' and the difficulty of stating the criterion for a single word in purely structural terms.

1.3. Let us return for a moment to the apparent analogy between IC analysis and functor analysis. A widely accepted formalization of IC structure is the so called PHRASE-STRUCTURE GRAMMAR (PSG).[11] As we have already seen, the segmentation of an expression by functor analysis does more than just assign words and phrases to various categories. It expresses relations EXPLICITLY that are not given directly in a PSG. Ajdukiewicz points out that the order of arguments may be used to show the subject-predicate relation, or the relation between antecedent and consequent in a hypothetical proposition. The fact that one part of an expression is the main functor while the other parts are its arguments may be used to show the relation of a modifier to the phrase which it modifies. In other words, the segments on each level (n^{th} order parts) have distinct roles in the whole expression, establishing a network of relations, and these roles are clearly indicated without any auxiliary devices beyond the fractional indices.

Not all linguists accept the PSG as a model for IC analysis. Gilbert Harman proposed a model in which additional gram-

9 For a discussion of zeroing, see Harris, *Mathematical Structures of Language*, 78-83.

10 Ajdukiewicz, "Syntactic Connexion", 213.

11 Chomsky, "On the notion 'rule of grammar'", 8-9.

matical information is supplied along with the category symbols. He writes:[12]

> ... we may describe a category by means of a basic category notation 'Noun Phrase' with several subscripts: 'Noun Phrase/Subject, Present Participle', where prior label indicates the basic category and the rest indicates that the noun phrase is the subject of a nearby present participle.

Harman replaces the category symbols used in PSG with category symbols followed by labels identifying the role of the phrases in the sentence. This brings the grammar closer to functor analysis since it does more than just segment and categorize. Chomsky refers to Harman's proposal as mere terminological equivocation:[13]

> This curious situation results simply from the author's redefinition of the term 'phrase structure' to refer to a system far richer than that to which the term 'phrase structure' has been universally applied in the rather ample literature on this subject.

1.4. The concept of SYNTACTIC CONNECTION which we discussed somewhat informally in section 1.2 is defined by Ajdukiewicz with the help of index sequences corresponding to word sequences.[14] The concepts leading up to this definition are summarized in (i) – (v) below.

Given an expression E:

(i) Permute the words of E so that a PROPER WORD SEQUENCE results. To do this, write down the main functor followed by its arguments in proper order (1st argument, 2nd argument, etc.). Repeat the same procedure with each of these first order parts (i.e. if any first order part is composite, write down its main functor followed by the arguments of that functor in proper order). Repeat with each 2nd order part, etc. until all composite parts have been rewritten and a sequence of single words results.

[12] Harman, "Generative grammars without transformational rules: a defense of phrase structure", 605.

[13] Chomsky, *Aspects of the Theory of Syntax*, 210.

[14] Ajdukiewicz, "Syntactic Connexion", 213-216.

This is the proper word sequence of the expression E which is now rewritten in PREFIX NOTATION or POLISH NOTATION. For example, (13) becomes:

and the lilac smells very strongly the rose blooms
and smells very strongly the lilac blooms the rose
and very strongly smells the lilac blooms the rose.

The final line is the proper word sequence, with each functor preceding its argument(s).

(ii) Write down the indices of the words in the same order as the words in the proper word sequence obtained in (i). This is called the PROPER INDEX SEQUENCE of the expression. Continuing with the above example, the proper index sequence is:

$$\frac{s}{ss} \quad \frac{\dfrac{\frac{s}{n}}{\frac{s}{n}}}{\dfrac{\frac{s}{n}}{\frac{s}{n}}} \quad \frac{\frac{s}{n}}{\frac{s}{n}} \quad \frac{s}{n} \quad \frac{n}{n} \quad n \quad \frac{s}{n} \quad \frac{n}{n} \quad n$$

(iii) Reading from left to right in the proper index sequence, look for the first fractional index which is followed by exactly the indices indicated in its denominator (in the same order). Replace this combination of indices by the numerator of the fractional index. (This amounts to finding a functor and all its arguments, thus forming a phrase whose category is given by the numerator of the fractional index of that functor.) The new index sequence is the 1st DERIVATIVE of the proper index sequence. E.g. the second and third indices in the proper index sequence above 'cancel', yielding the first derivative:

$$\frac{s}{ss} \quad \frac{\frac{s}{n}}{\frac{s}{n}} \quad \frac{s}{n} \quad \frac{n}{n} \quad n \quad \frac{s}{n} \quad \frac{n}{n} \quad n$$

(iv) To get each succeeding derivative scan the preceding one from left to right and perform one replacement (cancellation) as described in (iii). In the present example the result is that the 7th derivative consists of the single index s.

(v) The final derivative obtained by repeating this procedure until no further replacements (cancellations) are possible is called the EXPONENT of the original expression. The exponent in our example is s.

DEFINITION:[15] An expression is SYNTACTICALLY CONNECTED if and only if (1) it is well articulated throughout; (2) to every functor that occurs as a main functor of any order, there correspond exactly as many arguments as there are letters in the denominator of its index; (3) the expression possesses an exponent consisting of a single index.

The above example meets the criteria of this definition, hence is syntactically connected. Since the exponent is s the string belongs to the sentence category. Furthermore, within the string there are various syntactically connected substrings: *the rose* is syntactically connected and of category n; *very strongly* is syntactically connected and of category $\frac{s}{n}/\frac{s}{n}$; *strongly* is syntactically connected and of category $\frac{s}{n}/\frac{s}{n}$. Each word is syntactically connected and its category is, of course, given by the index assigned to it in the expression in which the word occurs.

It will be shown later that it is not necessary to rewrite the original expression in prefix notation (step (i)) in order to define the concepts of derivative and exponent. The 'reduction' to an exponent can be carried out by means of an algorithm working directly from the index sequence corresponding to the sequence of words in the original expression.

1.5. Ajdukiewicz makes a distinction between FUNCTORS and OPERATORS. He lists as operators the universal quantifier, existential quantifier, algebraic summation sign $\sum_{k=1}^{n}$, product sign $\prod_{x=1}^{n}$ and

[15] Ajdukiewicz, "Syntactic Connexion", 216.

definite integral sign $\int_a^b \ldots dx$. The chief distinction is that operators, as he uses the term, bind one or more variables. Functors are non-binding. Another difference is that a functor can be an argument of another functor; an operator cannot.

Indices may be assigned to operators as well as to functors provided we take into account the fact that an operator cannot be an argument. Recall that in a proper word sequence a functor always precedes its arguments. Using this same rule to deal with expressions containing operators, the operator index will be to the left of the indices of the operand expressions. Hence the operator index will not combine with any index on its left in a proper index sequence, since the operator will not be an argument. This will likewise be true in all derivatives of the proper index sequence. Ajdukiewicz uses a fraction with a vertical line on the left for an operator index. An operator is treated as a single word and receives a single index. For example,

(16)	$\supset$	$(\forall y)$	$(\exists x)$	f	x	y	p
proper index sequence:	$\frac{s}{ss}$	$\left\vert\frac{s}{s}\right.$	$\left\vert\frac{s}{s}\right.$	$\frac{s}{nn}$	n	n	s
1st derivative	$\frac{s}{ss}$	$\left\vert\frac{s}{s}\right.$	$\left\vert\frac{s}{s}\right.$	s			s
2nd derivative	$\frac{s}{ss}$	$\left\vert\frac{s}{s}\right.$	s				s
3rd derivative	$\frac{s}{ss}$	s					s
4th derivative	s						

(16) is syntactically connected since its exponent is a single index.

In addition to being syntactically connected, an expression containing operators must also meet the following condition in order to be SYNTACTICALLY CORRECT:[16]

... to each variable contained in the operator there must correspond, in the argument of the operator (i.e. in the expression to which the

[16] Ajdukiewicz, "Syntactic Connexion", 227.

operator applies) a variable of the same form which is not bound within the argument.

Ajdukiewicz's paper has been discussed here at considerable length since, historically, it is the basic work in this line of investigation from a linguist's point of view. However, I have omitted much material that is important from a logician's point of view, and Ajdukiewicz was addressing himself primarily to logicians. As a matter of fact, at the very outset he stresses the importance of linguistic syntax for logic and, in particular, the topic of syntactic connexion – especially in connection with the 'logical antinomies'.

1.6. In 1949, fourteen years after the publication of Ajdukiewicz's paper, an article titled *On The Syntactical Categories* appeared in *The New Schoolmen.* This article, written by I. M. Bocheński, O. P., had as its aim "to develop further the main ideas proposed by Professor Ajdukiewicz by drawing a sketch of such a theory and applying it to some logical and ontological problems" (p. 258). In a footnote he adds "... there is an ontological background in any language: Syntax mirrors Ontology".

Bocheński defines 'belonging to the same syntactic category in a language' in terms of mutual substitutability preserving well-formedness. In order to state the definition he first introduces four primitive terms, intuitively explained:[17]

Sy (x,L)	x is a SYMBOL of the language L
P (x,y,L)	x is a PART of y in L
Fl (x, L)	x is a WELL FORMED FORMULA of L
Sb (x,y,u,v)	v is a SUBSTITUTION of y for x in u

(To be a symbol in L, x must have an autonomous meaning in L; a wff in L is a symbol in L whose parts are arranged "according to the syntactical laws of L"; and Sb (x,y,u,v) if and only if v is like u except for containing y everywhere u contains x.)

DEFINITION: The symbols x, y belong to the same syntactic category of the language L if and only if, for every u and v if Sb (x,y,u,v) and Fl (u,L), then also Fl (v,L); and vice versa.

[17] Bocheński, "On The Syntactical Categories", 259.

This definition of 'belonging to the same syntactic category' is purely a syntactic one. The requirement is mutual substitution everywhere with preservation of well formedness.

Bocheński uses the term *operator* instead of *functor.* To define *operator* he introduces the idea of one symbol 'determining' another. As every linguist knows, a sentence is more than just a string of words; not only the words, but the relations between them must be understood in order to form a meaningful whole. In Bocheński's terminology, symbols must be connected by 'determination'. Thus he introduces another primitive term – DETERMINES. In the sentence *Bill disappeared* the word *disappeared* determines *Bill,* in *John likes apple pie* the word *likes* supplies the determination, and in the phrase *John and Bill* it is *and.*

DEFINITION:[18] The symbol x determines the symbol y if and only if what is meant by x is a property of what is meant by y – the word *property* being understood in the widest possible sense, which includes essential factors, necessary and accidental properties and also relations. For if R is the name of a relation which holds between what is symbolized by x and y, we shall say that R determines x and y.

X IS AN OPERATOR OF Y – Op (x,y) – if and only if x determines y.

When x is an operator of y, y is called the argument of x. An operator may have more than one argument; e.g. *loves* in *John loves Mary* has two arguments. This may be written *loves (John, Mary).* Also, *John loves Mary and Phyllis* may be written as:

(17) loves (John, and (Mary, Phyllis))

A single occurrence of a symbol cannot serve as an argument for two different operators. Thus in (17) *Mary* is an argument of *and,* but not of *loves.*

Bocheński's definition of syntactic category is purely syntactic, but the definition of operator is semantic. The primitive notion

18 Bocheński, "On The Syntactical Categories", 263.

'x determines y', which is the basis for the definition of operator, is a semantic relation. Bocheński confesses that the term *determines* is somewhat vague since no prior semantic system is given. To define operator syntactically would require the syntactic rules in advance – the very rules that operators are to help explicate.

Symbols which can occur as arguments, but not as operators, are called FUNDAMENTAL SYMBOLS; the syntactic categories of such symbols are called FUNDAMENTAL SYNTACTIC CATEGORIES – e.g., the category of names (n) and the sentence category (s). These are not the only possible fundamental categories. In a system of logic one may want to add a fundamental category of universal names. It is an open question as to what fundamental categories may be needed in a natural language besides n and s.

Bocheński discusses the syntactic categories used in *Principia Mathematica.* He shows that the logical antinomies result from a failure to consider the syntactic categories of the symbols, and from incorrect substitutions. When a symbol x operates on a symbol y, then x and y cannot be of the same syntactic category. (Note: A functor always has a fractional index. Let $\frac{a}{b}$ be the category of a given functor. If the argument of that functor also has the category $\frac{a}{b}$, then we get $\frac{a}{b}\frac{a}{b}$ in the index sequence. In this case no cancellation is possible. If $\frac{a}{b}$ is the index of the functor, then the argument of the functor must have an index b in order to form a phrase of category a with the functor. Obviously, $\frac{a}{b}$ cannot be identical with b.) In the antinomy concerning a property which is not a property of itself,

$$P(x) \equiv \sim (x(x))$$

and substitution of P for x gives:

$$P(P) \equiv \sim (P(P))$$

which results in a contradiction. If P is a property of P, it is not a property of P; and if it is not, then it is. But x(x) is not a well formed formula syntactically. The two occurrences of x belong to the same syntactic category so that one cannot be an argument of the other.

Regarding the common practice of taking s and n as the only two primitive categories, Bocheński reminds us that nothing in the theory of syntactic categories prohibits the introduction of new syntactic categories. To illustrate this point he considers the sentence:

(18) I wish to smoke

and suggests that such sentences contain another sentence "more or less explicitly". Thus (18) might be 'expanded' to:

(19) I wish that I smoke

He analyzes (19) by introducing a new syntactic category e, the category of 'enuntiables' – a term borrowed from Saint Thomas Aquinas:

(20) $$\underset{\frac{s}{ne}}{\text{wish}}\ \{\underset{n}{\text{I}},\ [\underset{\frac{e}{s}}{\text{that}}\,(\underset{\frac{s}{n}}{\text{smoke}}\,(\underset{n}{\text{I}}))]\}$$

In (20) *that* is an operator which forms an enuntiable *that I smoke* out of the sentence *I smoke*. The operator *wish* then forms a sentence out of two arguments, *I* and the enuntiable *that I smoke*.

Aside from the introduction of a new category, we see in this example the development of a transformational point of view. Like Ajdukiewicz, Bocheński sees the need to bolster functor analysis by relating certain sentences to other 'implicit' sentences. Later we shall also see how category assignments may take the place of certain transformations.

1.7. THE MEANING OF A FUNCTOR

The same functor index may be associated with functors that differ widely in meaning. In the sentences:

(21) Mary picked a daisy
(22) The men defoliated the forest
(23) A man who had been there wrote the article

the functors *picked*, *defoliated* and *wrote* all bear identical indices. But the index, which represents the category of the functor, does not give the MEANING. Rather it stands for a class of functors that have certain syntactic properties in common. Of course, we recognize that two functors of the same category are different by the very fact that their meanings differ; and since the functors are phrases in the language, their meanings are presumably known. But how does one, in principle, specify the meaning of a functor?

FUNCTOR is a relational concept. We may think of *picked* in (21) as a relation holding between *Mary* and *a daisy*; it is also a relation which holds between *Buford* and *the watermelon* or between *the migrant workers* and *cotton*. Since an n-place relation can be defined as a set of ordered n-tuples, it is tempting to define *picked* as the set of all ordered pairs (x,y) such that *x picked y* is an English sentence:

(24) picked = {(Mary, a daisy), (Buford, the watermelon),...}

A functor which takes only one argument might be thought of as a unary relation. E.g. *dream*, as it occurs in *men dream*, would then be defined by the set of all expressions *x* such that *x dream* is an English sentence. Words such as *if* or *depend* would not be in this set since *if dream* and *depend dream* are not English sentences.

Zellig Harris has suggested (in a conversation) that the meaning of a functor might be defined in this manner – by the set of acceptable arguments of the functor.[19] If so, then one could start with words that do not occur as functors, only as arguments, and use these words to get the meanings of certain functors. The latter could then be used to obtain the meanings of other functors, etc.

It is interesting to compare this concept of the meaning of a

[19] For a discussion of semantics in Harris's theory of transformations see his *Mathematical Structures of Language*, *211*.

functor with that of Bocheński. If x is a one argument functor and y is an argument of x, then in Bocheński's terminology, x is an operator of y. Let Mx and My be the meanings of x and y respectively: from Bocheński's definitions it follows that:

(25) Mx is a property of My

Now if properties are taken as classes (so that $a \in b$ renders *b is a property of a*), then:

(26) $$My \in Mx$$

This result is similar to Harris's suggestion. However, the statement that the meaning of a functor is given by the set of its acceptable arguments needs to be elaborated. When a phrase is substituted for *x* in *x dreams* the acceptability of the resulting phrase may be questionable. Instead of only two judgements, acceptable or not acceptable, there are relative degrees of acceptability leading to a partial ordering of the values of *x* in *x dreams*. In line with Harris's notion of an ACCEPTABILITY GRADING[20] we may say that the meaning of the functor *dream* is given by an acceptability grading over *x* in *x dreams*. This gives more information than an unordered set of values for *x*.

The definition of the meaning of a functor in terms of an acceptability grading over its arguments can be extended to functors of two or more arguments. E.g. the meaning of *picked* would be represented by an acceptability grading over *x* and *y* in *x picked y*. In general, we could say that two functors of the same category have the same meaning if and only if the acceptability grading over the arguments of one is the same as that over the arguments of the other.

1.8. THE HIERARCHY OF CATEGORIES

There is a discussion of functors and semantic categories in Tarski's *The Concept of Truth in Formalized Languages* (included in *Logic*,

[20] Harris, *Mathematical Structures of Language*, 53.

Semantics, Metamathematics, Oxford University Press). Tarski classifies semantic categories by assigning to a category, and to all expressions belonging to that category, a natural number called the ORDER of the category or expression (see p. 218, footnote 2, in the above mentioned book):

1st order: "sentences, names of individuals and expressions representing them"

(n+1)th order: "functors with an arbitrary number of arguments of order $\leqslant$ n, which together with these arguments form expressions of order $\leqslant$ n, but are not themselves expressions of the nth order" (At least one of the arguments must be of order n).

This definition does not include signs which bind variables; such signs (universal and existential quantifiers, the integration sign in calculus, etc.) are called OPERATORS. The distinction between finite order and infinite order languages plays a central role in the results of Tarski's book.

2

CATEGORIAL GRAMMARS FOR NATURAL LANGUAGES

2.1. The line of investigation from Husserl, Leśniewski, Ajdukiewicz and Bocheński was also pursued by the logician Y. Bar-Hillel. In 1950 a chapter from his doctoral thesis, revised and expanded, was published in the *Journal of Symbolic Logic* with the title *On Syntactic Categories*.[1] Around 1951 Bar-Hillel became involved in research on machine translation at M.I.T. In an article which appeared in *Language* 29 (1953), *A quasi-arithmetical notation for syntactic description*, he outlined a method for presenting the syntax of a natural language in a form which lends itself to computing the grammatical category of a string of words. He extended the method of Ajdukiewicz by (a) permitting a word to belong to more than one category, (b) allowing arguments to appear on the left of the corresponding functor as well as on the right and (c) introducing a cancellation rule to take care of (b).

In his 1950 article *On Syntactic Categories* Bar-Hillel begins by pointing out an important difference between constructed calculi and natural languages. Suppose P and Q are first level one-place predicates and a and b are individual symbols. In most constructed calculi if Pa and Qa are sentences, then if Pb is a sentence so is Qb. To show that natural languages do not have this nice property he uses an example given by Carnap. For P, Q, a and b use *is red*, *weighs five pounds*, *this stone* and *aluminum* respectively:

Pa : *This stone is red*
Pb: *Aluminum is red* (meaningful sentences)
Qa : *This stone weighs five pounds*
Qb: *Aluminum weighs five pounds* (not a meaningful sentence)

[1] Reprinted in *Language and Information*, 19-37.

Of course, this example presumes a certain point of view about sentences in a natural language that not all linguists agree with. Why not accept *Aluminum weighs five pounds* as a meaningful sentence which is simply false? As a matter of fact, Bar-Hillel notes that in a later article Carnap regards the sentence *This stone is now thinking about Vienna* as meaningful, but false.[2] Bar-Hillel's opinion about 'meaningless' strings of words is that "... in the verdict 'meaningless' ... is not merely a pragmatical statement of fact, it is a *decision*: He who gives this verdict declares by it that he is not ready to bother about the semantical properties of a certain word sequence and advises his hearers or readers to do the same."[3]

If one attempts to construct all and only the sentences of a natural language (taking this as a well defined set), then the rules of sentence formation presume some criterion for sentencehood other than empirical testing with native speakers. Any such criterion must lead to the same results as testing with native speakers if it is to be of any value. Bar-Hillel acknowledges the difficulty of constructing calculi which even closely approximate natural languages. He warns that the approximation of calculi to natural languages is a 'multidimensional' affair and that improvements might be made in one respect while falling short in others. Five model calculi are then presented and applied to a very small fragment of English. There is no attempt to approximate any natural language on a large scale.

In his preliminary definitions Bar-Hillel sets up a class of expressions (maximum genus) that corresponds to the concept of syntactic category. Two expressions belong to the same genus if and only if they are mutually substitutable in every sentence in a calculus, preserving sentencehood.

After becoming involved in research into machine translation, Bar-Hillel attempted to extend the ideas of Ajdukiewicz into a method of syntactic description for the whole of a natural language. The method is outlined in *A Quasi-Arithmetic Notation for*

[2] Carnap, "Testability and meaning", 5.

[3] Bar-Hillel, "On Syntactic Categories", in *Language and Information,* 35.

Syntactic Description (1953). Roughly, the idea is to assign each word to one or more categories in such a way that the syntactic description of any given string of words can be computed by means of the indices. This is, of course, basically the same plan followed by Ajdukiewicz, but with the prefix notation omitted and with cancellation on both the right and left permitted. As for the notation, the index for a string which with a string of category β on its immediate RIGHT forms a string of category α is written:

(1) $$\frac{\alpha}{[\beta]}$$

The index for a string which with a string of category β on its immediate LEFT forms a string of category α is written:

(2) $$\frac{\alpha}{(\beta)}$$

If an operator string forms a string of category γ out of m left arguments belonging to the categories $\alpha_1, \ldots, \alpha_m$ respectively and n right arguments belonging to the categories $\beta_1, \ldots, \beta_n$ respectively, then that operator string belongs to the category whose index is:

(3) $$\frac{\gamma}{(\alpha_1) \ldots (\alpha_m)\, [\beta_1] \ldots [\beta_n]}$$

For example,

(4) Sam fought the gooks

$$n \quad \frac{s}{(n)\,[n]} \quad \frac{n}{[n]} \quad n$$

1st derivative: $$n \quad \frac{s}{(n)\,[n]} \quad n$$

2nd derivative: s

The final derivative (exponent) is s, indicating that (4) is a sentence. There may be more than one derivation for a given index sequence:

(5)

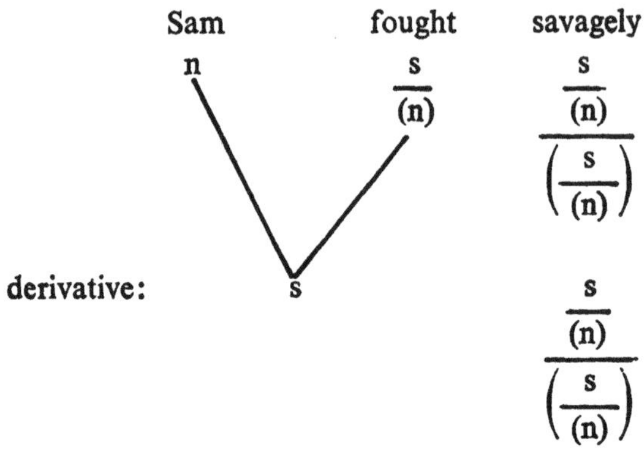

In this case no further cancellation is possible after the first derivative. However, there is another possible derivation from the same index sequence:

(6)

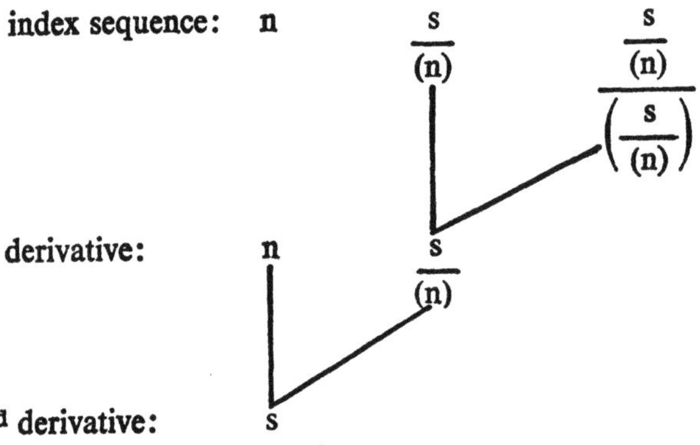

Thus *savagely* modifies the verb, not the sentence.

Any order of cancellation is permitted and different orders of cancellation yield different derivations. If in any of the derivations from a given index sequence the exponent is a single index, then the string with the given index sequence is said to be SYNTACTICALLY CONNEX and the resulting single index gives the grammatical category of the original string.

Each derivation tells something about possible constituent structures. (6) shows that *Sam* and *fought savagely* are the immediate constituents of the sentence. But the derivation (5) does not lead to a single index: *Sam fought* and *savagely* are not immediate constituents of the sentence. The index for *savagely* indicates that it modifies a verb, not a sentence, hence no cancellation is possible after the first derivative in (5). It follows that although *Sam fought* is syntactically connex in isolation, it is not connex in *Sam fought savagely*. Likewise, one could show that *people wept*, connex in isolation or within *people wept and dogs barked*, is not connex within *some people wept* or within *people wept bitterly*.

Connexity of a substring of a given string is defined as follows:[4]

a string m_1 is said to be connex at a certain place within a string m_2 with respect to the derivation d_1 if and only if:
(i) m_2 is connex
(ii) d_1 is proper (exponent is a single index)
(iii) d_1 includes a subderivation in which the index sequence of m_1 at the place in question has a proper exponent (single index).

Bar-Hillel interprets CONNEX WITHIN M_2 WITH RESPECT TO D_1 as CONSTITUENT OF M_2 WITH RESPECT TO D_1. However, there is actually more structural information in the former than in the latter for the reasons stated in 1.2 and 1.3.

Grammarians have long made a distinction between ENDOCENTRIC and EXOCENTRIC constructions. An endocentric construction has a constituent with (supposedly) the same distribution as the entire construction; all other constructions are exocentric. Of course, it is questionable whether two different strings can have IDENTICAL distributions. At any rate, assuming that the concept of endocentricity has been defined,[5] we may say that *poor John* is endocentric and that *John* is the head of this construction. On the other hand, *John sleeps* is exocentric.

Corresponding to the distinction endocentric-exocentric, Bar-Hillel classifies operators as ENDOTYPIC and EXOTYPIC: If an

[4] Bar-Hillel, "A Quasi-arithmetical Notation for Syntactic Description", in *Language and Information*, 67.
[5] See, e.g. Z. Harris, *String Analysis of Sentence Structure*, 12.

operator forms a string of the same category as its argument, the operator is endotypic (e.g. *blue* in *blue sky*, or *soundly* in *sleeps soundly*). Roughly, an endocentric construction corresponds to a phrase formed by an endotypic operator, and the argument of the operator is the HEAD of the construction.

The definition of endotypic operators may be extended to include operators of category $\alpha \ldots \alpha\backslash\alpha/\alpha \ldots \alpha$. For example, *and* in *roses are red and violets are blue* would then be considered endotypic since it is of category s\s/s.

In a later article (*Some Linguistic Obstacles to Machine Translation* (1960)) Bar-Hillel is less optimistic about the successful application of categorial grammars to natural languages. A number of difficulties are presented in that article which cast doubt on the adequacy of a grammar of the type proposed in *A Quasi-Arithmetic Notation for Syntactic Description.* Apparently there is a need for additional fundamental categories; the use of only n and s seemed inadequate to capture all the subtle distinctions in a natural language which lead to subcategorization. E.g. it is necessary to distinguish singular nominals from plural in English in order to avoid situations like:

bagels fattens
n n\s
s

Likewise there is a need to distinguish animate and inanimate, masculine and feminine. Another problem arises with articles and adjectives in English. If *the* belongs to n/n and *red* belongs to n/n, we have:

the red car
n/n n/n n
n/n n
n

but also:

red the car
n/n n/n n
n/n n
n

There is no claim that such problems are beyond solution, only that an undetermined number of fundamental categories are needed. As for the operator categories, there is a possibility that some words may have to be assigned to an infinite number of categories (e.g. *and*).

Another problem concerns the category of an element such as *up* in *look... up*, *call... up*, etc. An assignment which gives a proper derivation for *John looked up the word in the dictionary* fails for *John looked it up in the dictionary.*

If the index of an operator refers only to arguments on the immediate right and left of the operator, then it is difficult to find an index for a word such as *unfortunately*:

Unfortunately, the crowd was angry
The crowd, unfortunately, was angry
The crowd was anrgy, unfortunately

Presumably *unfortunately* is a sentence operator (with index s/s in the first example), but it may appear to the left, to the right or be inserted within its operand.

A different type of problem arises in the application of categorial grammars to machine translation. Since a word may belong to many different categories, a given sequence of words will have many possible index sequences. For each of these index sequences there may be more than one derivation. Suppose a machine is set up to scan each index sequence, perform all possible cancellations (in all possible orders) and thus perform all possible derivations corresponding to the given word sequence. This operation would, of course, reveal all the proper derivations. If k derivations lead to an exponent s, this reveals a k-fold ambiguous sentence. If there are only a few words in the given word sequence, the method is practical, but according to Bar-Hillel a sequence of thirty words could require trillions of machine operations. Sentences of thirty or forty words are not at all uncommon (e.g. the first three sentences in the article under discussion all exceed thirty words).

Bar-Hillel suggests that the use of a transformational model in place of an IC model would lead to a reduction in the number of categories to which a word must be assigned. The categorial grammar would suffice for the terminal strings underlying sentences of the kernel (referring to Chomsky's transformational model in *Syntactic Structures*); transformations would take care of the rest. A word need be assigned to only those categories required by the simpler set of strings. Furthermore, if a word does not occur in the set of strings underlying the kernel, that word belongs to no category! An example given of such a word is *sleeping* since it occurs only in sentences that result from transformations.

Bar-Hillel apparently accepts the notion that categorial grammars are inadequate for the description of natural languages on the grounds that they are 'equivalent' to PSG's as defined by Chomsky, and that PSG's have already been proven inadequate. The equivalence between categorial grammars and PSG's is shown in the paper *On Categorial and Phrase Structure Grammars*[6] in which Bar-Hillel collaborated with C. Gaifman and E. Shamir. Two points should be noted concerning the proofs:

(i) Equivalence of categorial grammars and PSG's refers to WEAK equivalence only. In other words, both types will 'generate' the same sets of strings. It does not follow that they give the same information about the structure of those strings (see pages 17-20 of this book).

(ii) The weak equivalence referred to in (i) may not hold for all types of categorial grammars, only for those similar to Bar-Hillel's. (I will propose an extension of categorial grammars for which even this weak equivalence may not hold.)

2.2. In a paper titled *The Mathematics of Sentence Structure*[7] Joachim Lambeck outlined a theory of syntactic types with the stated purpose of obtaining an algorithm for distinguishing sentences from non-sentences in a natural language (or fragments of a natural language). The method is similar to the ones already

[6] Reprinted in *Language and Information*, 99-115.

[7] *American Mathematical Monthly* 65, No. 3, 154-170.

discussed and the term *type* is used instead of *category*. But there are some departures from the works so far discussed which are of interest.

Like his predecessors, Lambek takes as primitive the type s of sentences and the type n of names. He restricts *sentence* to "complete declarative sentences". The TYPE n IS ASSIGNED TO ALL EXPRESSIONS WHICH CAN OCCUR IN ANY CONTEXT IN WHICH ALL PROPER NAMES CAN OCCUR. This implies that type n is not assigned to count nouns. For example, *book* cannot replace *John* in *John is here*. Neither is type n assigned to pronouns since we have *Big John is here* but not *Big he is here*. But mass nouns such as *milk* and *rice* and noun phrases such as *poor John* and *fresh milk* are assigned type n. Supposedly these mass nouns and noun phrases can occur in any context in which all proper names can occur – a difficult hypothesis to verify.

In order to avoid violating number agreement in cases like:

men	works
n	n\s

and

John	work
n	n\s

Lambek suggests adding another primitive type n* for noun plurals. The index sequence for *men work* then reduces to s while that for *men works*, does not:

men	work		men	works
n*	n*\s		n*	n\s
s				

This solution to the problem of number agreement also entails the assignment of an additional type to adjectives as well as to nouns and verbs:

poor	John		poor	people
n/n	n		n*/n*	n*
n			n*	

Although troublesome in English, this would work well in Spanish or French where adjectives are marked for number.

Since pronouns cannot be assigned type n, they are given compound types that distinguish subject position from object position. The assignment of $s/(n\backslash s)$ to *he* yields:

$$\begin{array}{cc} \text{he} & \text{sleeps} \\ s/(n\backslash s) & n\backslash s \\ s & \end{array}$$

and

$$\begin{array}{ccc} \text{poor} & \text{he} & \text{sleeps} \\ n/n & s/(n\backslash s) & n\backslash s \\ n/n & s & \end{array}$$

with no further cancellation possible in the latter. The pronoun *him* is assigned type $(s/n)\backslash s$:

$$\begin{array}{ccc} \text{Jane} & \text{likes} & \text{him} \\ n & n\backslash(s/n) & (s/n)\backslash s \\ s/n & & (s/n)\backslash s \\ s & & \end{array}$$

These type assignments do not suffice for sentences such as *she likes him.* Neither:

$$\begin{array}{ccc} \text{she} & \text{likes} & \text{him} \\ s/(n\backslash s) & n\backslash(s/n) & (s/n)\backslash s \end{array} \quad \text{nor} \quad \begin{array}{ccc} \text{she} & \text{likes} & \text{him} \\ s/(n\backslash s) & (n\backslash s)/n & (s/n)\backslash s \end{array}$$

can be further simplified by cancellation. To handle such situations Lambek includes the rules:

$$(x\backslash y)(y\backslash z) \rightarrow x\backslash z \qquad \text{and} \qquad (x/y)(y/z) \rightarrow x/z$$

(This last rule is due to Ajdukiewicz.) These rules permit either derivation:

$$\begin{array}{lll} s/(n\backslash s) & n\backslash(s/n) & (s/n)\backslash s \\ s/(n\backslash s) & n\backslash s & \\ s & & \end{array} \quad \text{or} \quad \begin{array}{lll} s/(n\backslash s) & (n\backslash s)/n & (s/n)\backslash s \\ s/n & & (s/n)\backslash s \\ s & & \end{array}$$

Lambek's syntactic calculus takes the following form:[8]

(i) An EXPRESSION is a string of words
(ii) Certain expressions are assigned PRIMITIVE types
(iii) If A has type x and B has type y, then AB has type xy
(iv) $x \rightarrow y$ means that any expression of type x also has type y
(v) If x and y are types, so are $y\backslash x$ and x/y

axiom schemes:

(a) $x \rightarrow x$
(b) $(xy)z \rightarrow x(yz)$ (b′) $x(yz) \rightarrow (xy)z$

rules of inference:

(c) if $xy \rightarrow z$ then $x \rightarrow z/y$ (c′) if $xy \rightarrow z$ then $y \rightarrow x\backslash z$
(d) if $x \rightarrow z/y$ then $xy \rightarrow z$ (d′) if $y \rightarrow x\backslash z$ then $xy \rightarrow z$
(e) if $x \rightarrow y$ and $y \rightarrow z$ then $x \rightarrow z$

According to (iii) xy is taken not just as a sequence of types, but as a type. It follows that a string may be said to have a type xyz...w even though xyz...w does not reduce to anything simpler. A string of words need not be a 'constituent' to be assigned a type.

Because of (b) and (b′) this system is called the ASSOCIATIVE calculus of types. One consequence of associativity is (7):

(7) $(x\backslash y)/z \rightarrow x\backslash(y/z)$ and $x\backslash(y/z) \rightarrow (x\backslash y)/z$

Lambek simply writes $x\backslash y/z$ for either grouping; e.g.:

John	likes	Jane
n	n\s/n	n

yields either:

n	[(n\s)/n	n]	or	[n	n\(s/n)]	n
n	n\s			s/n		n
s				s		

[8] Lambek, "The Mathematics of Sentence Structure", 163.

Another consequence of associativity is less palatable. If (b) is applied to the index sequence of the noun phrase *very cold water* the result is:

(A)	very	cold	water	(B)	very	cold	water
	((n/n)/(n/n)	n/n)	n	$\rightarrow$	(n/n)/(n/n)	(n/n	n)

This would seem to place *(very cold) water* on a par with *very (cold water)*. But only the grouping in (A) corresponds to a derivation which yields the exponent n. The first derivative in a derivation corresponding to (B) is ((n/n)/(n/n) n) which cannot be further reduced. This agrees with the generally held view that *very* modifies *cold*, not *cold water*.

The rules of the syntactic calculus permit 'expansion' of simple type symbols to compound type symbols; e.g.:

(8) $x \rightarrow (xy)/y$

proof: $xy \rightarrow xy$ by (a)

$\therefore\ x \rightarrow (xy)/y$ by (c)

Such expansions could lead to very complex type assignments. Lambek shows, however, that there is an effective procedure to determine whether a given formula $x \rightarrow y$ can be deduced from (a) – (e).

In a later paper "On the calculus of syntactic types"[9] Lambek abandoned the associative rules (b), (b′). He states that in the earlier paper many 'pseudo sentences' resulted from the assignment of types to unstructured strings and that types should only be assigned to PHRASES (bracketed strings). The NON-ASSOCIATIVE CALCULUS takes the following form:[10]

(i) all atomic phrases are phrases
(ii) If A and B are phrases, so is (AB)
(The atomic phrases are not identified.)
(iii) all primitive types are types
(iv) If x and y are types, so are (xy), (x/y) and (x\y)

9 In *Structure of Language and Its Mathematical Aspects*, ed. by R. Jakobson, 166-178.

10 Lambek, "On the calculus of syntactic types", 168.

Rules for assigning types to phrases:

(a) If A has type a and B has type b then (AB) has type (ab)
(b) If (AB) has type c for all B of type b then A has type (c/b)
(c) If (AB) has type c for all A of type a then B has type (a\c)

The rules (x\y)/z → x\(y/z) and (x/y)(y/z) → x/z which held in the associative calculus are not valid in this system. According to Lambek, the decision procedure for the associative calculus can be adapted for the non-associative calculus. Mechanical parsing of a string begins with bracketing of the string and assignment of types to the words in the string. Of course, there may be more than one way of bracketing the string and there are usually many possibilities for assigning a type from the 'dictionary' to a given word in the string. The desired goal is to OBTAIN ALL GRAMMAR RULES BY TYPE ASSIGNMENTS IN THE DICTIONARY.

Certain type assignments are shown to be equivalent to transformational rules. The assignment of s/(n\s) to *he* is equivalent to the transformational rule:

(9) If *nX* is a sentence then *He X* is a sentence.

Similarly, the assignment of (?)/(n\s) to *who* is equivalent to the transformational rule:

(10) If *n X* is a sentence then *who X* is a sentence (interrogative).

There is no claim that ALL transformational rules can be replaced by type assignments. In particular, 'elliptical' transformational rules may not be replaceable by type assignments within the present framework.

2.3. EQUIVALENCE OF CATEGORIAL GRAMMARS OF BAR-HILLEL AND LAMBEK

The systems of both Bar-Hillel and Lambek are designed to provide for the mechanical parsing of arbitrary strings of words in a natural language, leading to the determination of the grammatical category of each string. We have seen that their methods differ in certain respects; the question then is whether these differences are really substantial. Joel M. Cohen proved[11] that the categorial grammars of Bar-Hillel and Lambek are equivalent in the following sense:

Let G be a categorial grammar and L(G) the set of all strings that cancel to s under the rules of G; then L(G) is called the LANGUAGE OF G. If G_1 and G_2 are categorial grammars of Bar-Hillel and Lambek respectively, then $L(G_1) = L(G_2)$; i.e. with a given vocabulary V, the set of strings over V which one grammar classifies as sentences is identical with the set of strings over V which the other grammar classifies as sentences.

In an article *On Categorial and Phrase Structure Grammars* (1960) Bar-Hillel showed that unidirectional, bidirectional and restricted categorial grammars are all weakly equivalent (and also that these are weakly equivalent to PSG). A UNIDIRECTIONAL categorial grammar uses only right cancellation (a/b b → a) or only left cancellation (b b\a → a). The grammar of Ajdukiewicz is unidirectional since it uses only right cancellation. A BIDIRECTIONAL categorial grammar uses both right and left cancellation. In a RESTRICTED categorial grammar there are a finite number of primitive categories $r_1, ..., r_n$ and all operator categories are of the form $r_i \backslash r_j$ and $(r_i \backslash r_j) \backslash r_k$ (alternatively, r_i / r_j and $r_i/(r_j/r_k)$).

Cohen refers to the grammar of Lambek (1958) as a FREE CATEGORIAL GRAMMAR, abbreviated f.c.g., and that of Bar-Hillel simply as CATEGORIAL GRAMMAR, abbreviated c.g. An f.c.g. has, in addition to the rules (11):

(11) x/y y → x and x x\y → y

also the rules (12):

(12) x\(y/z) → (x\y)/z (x\y)/z → x\(y/z)
x/y y/z → x/z x\y y\z → x\z
x → y/(x\y) x → (y/x)\y

11 Cohen, "The Equivalence of Two Concepts of Categorial Grammar", 475-484.

The equivalence of c.g. and f.c.g. means that the strings which are accepted as sentences using (11) and (12) are also accepted as sentences by using (11) alone. This depends on the manner in which categories are assigned to elements in the vocabulary.

The method of Cohen's proof consists in showing that:

(i) given any f.c.g. there is a weakly equivalent bidirectional categorial grammar, and
(ii) given any bidirectional categorial grammar there is a weakly equivalent f.c.g.

The proof makes use of the fact that for any bidirectional grammar there is a weakly equivalent restricted one, and vice versa, already proved by Bar-Hillel. E.g., given a bidirectional grammar G_b, the existence of a weakly equivalent restricted categorial grammar G_r is assured and (ii) can then be proved by producing an f.c.g. weakly equivalent to G_r.

A critical point in proving the equivalence of g.c. and f.c.g. is to show that the extra rules of cancellation in an f.c.g. which are not given in a c.g. can be circumvented by defining an appropriate ASSIGNMENT FUNCTION for the f.c.g. (The assignment function is that function by which a finite number of categories are assigned to each element in the vocabulary of a particular grammar.) Suppose an f.c.g. is given with an assignment function U. A new assignment function U′ may be defined such that for any element A in the vocabulary of the f.c.g.,

$U(A) \subset U'(A)$ and
if $y \in U'(A)$, there is some $x \in U(A)$ such that $x \rightarrow y$.

With this new assignment function U′ it turns out that the rules for a c.g. suffice to accept the same strings as sentences that were accepted by the f.c.g. with the function U. The details of Cohen's proof are rather sticky, and the reader is advised to consult the original article (see footnote 11).

The weak equivalence of two categorial grammars means only that they confer sentencehood on the same strings. As for the concept of *sentence*, a sentence is any string one of whose index sequences cancels to s. To be sure, the category or type assignments are made so that the outcome will conform with generally accepted feelings about what is or is not a sentence. But once these assignments are made on the basis of certain examples, the grammar blindly labels as a sentence any string with an index sequence that cancels to s. Thus Lambek's assignment of s\s to *here*[12] gives a perfectly acceptable result in:

```
John works  here
 n     n\s   s\s
 └─────┘
 s           s\s
 s
```

but it also yields:

```
John works here here
 n    n\s   s\s  s\s
 └─────┘
 s          s\s  s\s
 └──────────┘
 s               s\s
 └───────────────┘
 s
```

It would seem that some restrictions on context might be needed in assigning categories to words in such cases.

Finally, note that these grammars will decide for a given string of words in a finite number of steps either that the string is a sentence or that it is not a sentence. There is no built-in detector of degrees of grammaticality – only a recognition of the grammatical category of the string, if any.

2.4. Henry Hiż has dealt with the subject of grammatical categories and the use of functors in linguistics and logic in a variety of papers. In *The Intuitions of Grammatical Categories*[13] he discusses three important factors that influence the grouping of linguistic segments into grammatical categories: (1) intersubstitutability, (2) structural connectivity and (3) roles in transformations.

[12] Lambek, "The Mathematics of Sentence Structure", 156.
[13] Methodus (1960), 311-319.

Structural connectivity refers to the relations between parts of a sentence, some parts being treated as functors and others as arguments of those functors. More precisely,[14]

a grammatical category as a component in a structure of a sentence may be viewed as a three place relation between a resulting grammatical category and two sequences of grammatical categories, namely, its left-hand sequence and its right-hand sequence.

Suppose that in a given sentence the string a consisting of $b_1 \ldots b_k\ x\ c_1 \ldots c_m$ occurs; the category of each b_i is known (say β_i) and the category of each c_i is known (say γ_i); and the entire string is of category α. This is shown schematically in (13), where the category of each segment is given below the segment:

(13) $$\begin{array}{ll} a = & b_1 \ldots b_k\ \ x\ \ c_1 \ldots c_m \\ \alpha & \beta_1 \ldots \beta_k \text{——} \gamma_1 \ldots \gamma_m \end{array}$$

The grammatical category of the segment x (as x occurs in the given sentence) is then written:

(14) $$(\alpha;\ \beta_1 \ldots \beta_k \text{——} \gamma_1 \ldots \gamma_m)$$

Comparing (14) with previous notation:

n\s/n = (s;n_n)	s/n = (s;_n)
n\s = (s;n_)	(s/n)/n = ((s;_n);_n)

(14) gives a good picture of the position of the functor x with respect to its arguments: the position of the dash '——' with respect to the left-hand and right-hand sequences of category symbols corresponds to the position of x with respect to its left arguments and right arguments respectively. This notation proves very useful for representing the grammatical categories of DISCONTINUOUS PHRASES; e.g., the category symbol for *if* ... *then* ... is (s;_s_s). (For a detailed account of the use of this notation and a rigorous definition of the cancellation procedure see *Grammar Logicism* by H. Hiż in *The Monist* 51 [1967], No. 1.)

The three criteria mentioned above for establishing grammatical categories do not necessarily lead to the same classification

[14] Hiż, "The Intuitions of Grammatical Categories", 312.

of segments in a language. It is not a question of which is the correct approach – each one answers a certain type of question about sentence structures: (1) what segments can replace one another in any sentence, preserving sentencehood? (2) What segments can occupy a certain position with respect to other segments of stated categories in a sentence? (3) What classes of segments are useful in the statement of transformational rules relating sentences of different forms?

The use of grammatical categories to show structural connectivity is related to definitions in formal systems: a definition should be such that the grammatical category of the definiendum can be computed from the grammatical categories of the parts of the definiens, and a constant symbol of a new grammatical category may be defined (i.e. when the category of the constant is computed from the other known categories in the definition, the category of the constant may be different from any that has occurred in the development of the system up to that point).

2.5. Given a sequence of linguistic segments, one may ask how this sequence can be 'completed' by the insertion of other segments to form new phrases. The fact that a segment x is of grammatical category:

$$(\alpha; \beta_1 \ldots \beta_k \text{——} \gamma_1 \ldots \gamma_m)$$

tells us that the insertion of x between b_k and c_1 in $b_1 \ldots b_k$ $c_1 \ldots c_m$ will COMPLETE that sequence to a string $b_1 \ldots b_k$ x $c_1 \ldots c_m$ of category α. To say that *the* is of category (n;_n) is to say that *the*, inserted before a segment of category n, completes that segment to a noun phrase: *the house*, *the good old days*, etc.

It seems that syntactic completion analysis and functor analysis are two sides of the same coin. Hiż develops the relation between the two points of view in the paper *Syntactic Completion Analysis and Theories of Grammatical Categories*.[15] He suggests that in the preceding example (*the house*) we may also want to consider

15 *Transformations and Discourse Analysis Papers* 21, University of Pennsylvania.

house as a segment which completes *the* to form a noun phrase. In this case, if we let Q be the category of *the* then *house* is of category (n;Q_). (Thus the article may be thought of either as 'noun determiner' or as 'noun determined'.)

Several interesting problems are presented in the paper on syntactic completion analysis along with some possible solutions which lie outside the framework discussed so far. Consider, e.g., the analysis of the sentence *To a man who was poor John gave money*.[16] One analysis would assign the category (S:PNN_N) to *gave*:

(15)	To	a man who was poor	John	gave	money
	P	N	N	(S; PNN_N)	N

Now if *was* is assigned (S; N_A), the result is:

(16)	... a man who	was	poor ...
	N	(S; N_A)	A

Next consider the category of *who*. With an N on its left *(a man)* and an (S;N_A)A on its right *(was poor)*, it forms a phrase of category N:

(17)	a man	who	was	poor
	N	(N;N_(S;N_A)A)	(S;N_A)	A

This entire analysis is shown in (18) where numerals are placed below the category symbols to show which arguments go with which functors (consistent with (15) – (17)).

(18)	To	a	man	who	was	poor	John	
	P	(N;_N)	N	(N;N_(S;N_A)A)	(S;N_A)	A	N	
	6	2 1	1	5 2 3 4	3	4	7	
	gave	money						
	(S;PNN_N)	N						
	9 6 5 7 8	8						

[16] Hiż, "Syntactic Completion Analysis", 24-25.

One question remains: Which symbols cancel with N and A in (S;N_A) below *was*? The A can be cancelled with the A under *poor*, but the N must be cancelled either with 2 (under *a man*) or with 5 (under *who*). However, 2 and 5 have already been cancelled. Hiż notes the possibility of permitting *a man* to be cancelled twice, once with 5 under *gave* and once with the N under *was*. His final conclusion is that the notation needs to be expanded to take into account the ENVIRONMENT in which a functor occurs, and that *who was poor* should be assigned to category S. The category assignment for *who* is then:

(19) (S;_(S;N_A)A) *if the first parenthetical expression to the left of the functor is of category N.*

In order to avoid such lengthy statements of environmental restrictions Hiż uses '$\_\_\__{-i}$' for the ith parenthetical expression to the left of the functor-argument cluster and '$\_\_\__{+i}$' for the ith parenthetical expression to the right:

(20) ... a man who [if $\_\_\__{-1}$ is N] was poor

a man	who		was	poor
N	(S;_(S;N_A)A)	[if $\_\_\__{-1}$ is N]	(S;N_A)	A
2	3 4		3	4

The environmental restriction is fulfilled by *a man*, but there is no longer an N under *who* which cancels with the N under *a man*. This leaves the N under *a man* available for further cancellation. *who was poor* is now analyzed as a sentence embedded within a sentence. In (20) *who* is given a CONTEXT SENSITIVE CATEGORY ASSIGNMENT.

The structural connectivity of the entire sentence may be pictured by connecting each functor to its arguments with lines (analyzing *who* as in (20)):

(21) To (a man) who was poor John gave money

(I have connected *a man* to *gave* since *who* is no longer N-forming as it was in (18). There is still a problem here with *was*.)

In (21) the connection between the embedded sentence *who was poor* and the sentence in which it is embedded, *To a man ...... John gave money*, is not shown by any of the connecting lines; it is not given directly by the functor-argument relation. There is no functor in either sentence with an argument in the other. The connecting link is given only by the statement on environmental restriction: [if$____{-1}$ is N]. Since this is not a functor-argument relation, it could be shown by a broken line as in (22).

(22) To (a man) who was poor John gave money

In (22) *a man* is not cancelled twice (in the usual sense) – only via the environmental condition. But no matter what device is used, there is still a structural connection shown between the embedded sentence and the rest. The introduction of an environmental condition appears to be a case of adding a new type of cancellation to the usual one. If so, we may ask whether having two types of cancellation is more desirable than having only one type and using it to cancel the same segment twice.

A functor whose index has no context sensitive statement attached refers to its environment none the less. The index says, in effect, that this phrase (the functor) in such and such an environment forms a new phrase together with that environment. An index with the environmental condition attached, as in (20), says: this phrase (the functor) in such and such an environment forms a new phrase together with that environment – provided there is an additional environment of so and so. This amounts to stating part of the necessary environment in one form and the remainder in another form.

Hiż's notation for the environmental condition permits reference to the environment at some distance from the functor-argument cluster. The symbol '$\_\_\__{k}$' refers to a parenthetical expression that is not contiguous with the functor-argument cluster whenever $|k| > 1$. Whether values of k other than -1, 0 and $+1$ are needed remains to be seen.

In *Computable and Uncomputable Elements of Syntax*[17] Hiż makes use of discontinuous functors. The procedure for cancellation which he gives in *Grammar Logicism* does not cover index sequences with discontinuous functor indices. In order to extend the cancellation rules to cover these cases some procedure such as the one illustrated below might be used.

The discontinuous functors *if ... then ..., either ... or ..., both ... and ...* have category symbols of the form (c;_a_b). Whenever a functor (f_1 ... f_2 ...) occurs in a string x, write the category symbol under f_1 only and write 0 under f_2. If x contains a substring $f_1 y f_2 z$ structured as in (23)

(23)

f_1	y	f_2	z
(c;_a_b)	a	0	b
3 1 2	1		2

then replace (23) by

(24)

f_1	y	f_2	z
c	*	*	*
3			

where the asterisks indicate the scope of c. E.g. consider the string *either p or if q then r* structured as in (25):

(25)

either	p	or	if	q	then	r
(S;_S_S)	S	0	(S;_S_S)	S	0	S
5 4 3	4		3 1 2	2		1

The immediate reduct of (25) is:

(26)

either	p	or	if	q	then	r
(S;_S_S)	S	0	S	*	*	*
5 4 3	4		3			

and the reduct of (26) is:

(27)

either	p	or	if	q	then	r
S	*	*	*	*	*	*
5						

[17] In *Logic, Methodology and Philosophy of Sciences III*, ed. by Rootselaar and Staal, 239-254.

This procedure may be generalized to include discontinuous functors of the form $f_1 \ldots f_2 \ldots f_n$ where $n \geq 2$.

In several papers Hiż stresses the idea that a text (sentence, etc.) has not just one structure, but many. We have already encountered this idea on pages 48-49 where it was pointed out that even a simple noun phrase *the house* could be assigned structure in more than one way. But if this is the case, how does one decide what structures to assign to a given text? Hiż writes:[18]

> The applicable operations determine which structures are to be assigned to the text. Some of the operations are close paraphrases, others are semantic changes, still others are drawing consequences.

2.6. It is generally accepted in papers on categorial grammars that a single word may occur in more than one grammatical category. Of course, one may argue that what appears to be one word occurring in two different categories is actually two different words which are homonymous. Such an argument may have appeal in:

(28) I enjoyed the HIKE. Let's HIKE to the top of the ridge

but not so much in:

(29) These roses are RED. These are RED roses.

In logic a constant symbol can be used in different grammatical categories. Hiż makes this point in a paper "On the Abstractness of Individuals".[19] He gives the example of ⊃ in:

(30)
$$\begin{array}{lllll} \bigwedge x \bigwedge y \ulcorner x & & \supset & y & \equiv \\ & (S;\_S) & (S;(S;\_S)\_(S;\_S)) & (S;\_S) & \end{array}$$

$$\begin{array}{lllll} \bigwedge z \ulcorner x & (z) & \supset & y & (z)\urcorner\urcorner \\ (S;\_S) & S & (S;S\_S) & (S;\_S) & S \end{array}$$

As for variables, it is customary to take all occurrences of a variable that are bound by the same quantifier to be of the same

[18] Hiż, "Computable and Uncomputable Elements of Syntax", 243.
[19] In *Identity and Individuation*, 251-261.

grammatical category (i.e. spectrum = 1). But in the above mentioned paper different occurrences of a variable bound by the same quantifier are taken in different categories; e.g. the two occurrences of f inside the corners in (31):

$$
\begin{array}{llllllll}
(31)\ \wedge f\,\ulcorner & f & x & y & \equiv & x & f & y\urcorner \\
 & ((S;\_a);\_a) & a & a & & a & (S;a\_a) & a \\
 & \quad 2 \quad\ 1 & 1 & 2 & & 3 & \quad 3\ \ 4 & 4
\end{array}
$$

In effect, (31) says that any two-argument functor in the infix notation is equivalent to another functor expressed in the prefix notation. One of the results obtained by permitting a spectrum greater than 1 is a solution to RUSSELL'S ANTINOMY. If x has a spectrum of 1, then the antinomy may be stated as follows:

$$(32)\quad \wedge x\,\ulcorner R(x) \equiv \sim (x(x))\urcorner \qquad (R(x) \text{ may be read } x \in R)$$

Substitute R for x in (32)

$$(33)\qquad\qquad R(R) \equiv \sim (R(R))$$

and the result is a contradiction. But if the spectrum of x is greater than 1, then (32) may be replaced by either:

$$
\begin{array}{llllll}
(34)\ \wedge x\,\ulcorner & R & (x) & \equiv \sim & (x & (x))\urcorner \\
 & (S;\_a) & a & & (S;\_a) & a
\end{array}
$$

or:

$$
\begin{array}{llllll}
(35)\ \wedge x\,\ulcorner & R & (x) & \equiv \sim & (x & (x)) \\
 & (S;\_(S;\_a)) & (S;\_a) & & (S;\_a) & a
\end{array}
$$

The substitution of R for x in (34) gives:

$$
\begin{array}{lllll}
(36)\ & R & (R) & \equiv \sim\ (R & (R)) \\
 & (S;\_a) & a & (S;\_a) & a
\end{array}
$$

and a contradiction still follows. However, substitution of R for x in (35) gives:

$$
\begin{array}{lllll}
(37)\ & R & (R) & \equiv \sim\ (R & (R)) \\
 & (S;\_(S;\_a)) & (S;\_a) & (S;\_a) & a
\end{array}
$$

and no contradiction follows since the grammatical analysis of R(R) on the right side of (37) differs from that on the left. (37) does not say that a proposition is equivalent to its own negation, but that a proposition p is equivalent to the negation of another proposition q.

As we saw in section 2.4, a definition may introduce a new grammatical category in the definiendum. This is the case in (35). Hiż adds the following criterion to the rule of definition:[20]

> When in doubt, the grammatical category of a variable in the definiendum should be taken as the highest in which this variable occurs as free in the definiens.

Thus in (35) the grammatical category of x in R(x) is taken as (S;_a) since this is the highest category in which the variable x appears in ~ (x(x)). By using this criterion Russell's antinomy is avoided.

2.7. A slightly different functor notation is used by Haskel B. Curry in a paper titled *Some Logical Aspects of Grammatical Structure.*[21] He uses blanks to indicate where the arguments go and subscripts to indicate first argument, second argument, etc. Some examples are:

(i) red__$_{1}$	(as in *red rose*)
(ii) __$_{1}$ melts	(as in *ice melts*)
(iii) __$_{1}$ melts __$_{2}$	(as in *fire melts ice*)
(iv) both__$_{1}$ and__$_{2}$	(as in *both John and Mary*)

Functors are classified by the number and type of arguments and the type of phrase that results. The notation is FAB ... C where A is the category of the first argument, B the category of the second argument, ..., and C the category of the resulting

[20] This means that when the difference is in the ORDER of categories in the Tarski sense (see 1.8) then the category of highest order is taken into the definiendum.

[21] In *Structure of Language and Its Mathematical Aspects*, ed. by R. Jakobson, 56-68.

phrase. Thus (i), (ii), (iii) and (iv) are of categories FNN, FNS, FNNS and FNNN respectively.

Curry takes a broad view of functors. He writes:[22]

What Harris and Chomsky call transformations are also functors. A functor is *any* kind of linguistic device which operates on one or more phrases (the argument(s)) to form another phrase.

2.8. This raises a question of terminology: whether to consider a functor as any kind of operator on phrases in a language to form another phrase in the language, or to restrict the term *functor* so that a functor must itself be a phrase in the language. Most of Harris's transformational operators are functors in this restricted sense: e.g. *and, or, begin, have-en, I know that,* etc. Others, such as φ_p (permutation), are not.

In the following chapters I shall use the term *functor* in the restricted sense: functors and their arguments will consist of phrases in the language under study. A phrase may be a word, morpheme or combination of these – including suprasegmental morphemes. This interpretation of *functor* is in keeping with the point of view that the phrases which make up an expression should provide the basis for a syntactic analysis of that expression. The policy adopted here is to pursue this point of view as far as practical.

Functors in the restricted sense may take the place of certain transformations, as has already been shown. This does not rule out the use of transformations, but casts them in a different role.

[22] Curry, "Some Logical Aspects of Grammatical Structure", *62.*

3

GRAPHS AND STRUCTURED STRINGS

3.1. A certain point of view about natural languages has pervaded linguistics in recent years. It is clearly stated in Chomsky and Miller's *Introduction to the Formal Analysis of Natural Languages*:[1]

> We consider a language to be a set (finite or infinite) of sentences, each finite in length and constructed by concatenation out of a finite set of elements.

Thus grammatical analysis is focused on the sentence, and a sentence is a finite string over a finite alphabet. A grammar is then taken to be a set of rules that (recursively) specify these sentences. Theories of language that follow this line of thought make a basic assumption that a language is a WELL DEFINED set of strings and a grammar of the language is a 'machine' that cranks out exactly those strings. In this string-machine approach structure is assigned to a given string by the very process of cranking out the string.

The approach of the present paper differs from the above in several respects. Analysis will not be limited to sentences, but will include longer texts as well (a TEXT may include one or more sentences). No program will be offered for deriving all the sentences or texts of a language from a given subset of sentences or texts, nor from other more abstract objects. The emphasis will be on RELATIONS between phrases – relations between components of a text or between different texts. These relations determine the

[1] In *Handbook of Mathematical Psychology*, vol. 2, 283-284. Also stated in *Syntactic Structures* (p. 13) omitting the phrase "by concatenation".

relevant structures that may be assigned to a text. The spirit of this investigation is to try to characterize the structure of a text principally in terms of the phrases that actually occur in the text and the relations between these phrases. The basic tool of analysis will be the representation of the phrases within a text in terms of functors and their arguments. Directed graphs will be utilized as a notational device better suited than strings for direct representation of relations within a text. Rules will be given later for constructing graphs corresponding to the assignment of functors to the phrases of a text.

The non-stringlike nature of a text becomes apparent when one considers suprasegmentals such as intonation patterns, emphatic stress and contrastive stress occurring simultaneously with segmental elements. The existence of discontinuous morphemes poses another problem for anyone who considers a sentence as a string (presumably a concatenation of morphemes or formatives of some kind). The complex relations between various parts of a sentence are usually stated in terms of a DERIVATION of the sentence, and this derivation consists of a sequence of strings whose elements may be abstract symbols, actual phrases or sets of features. In this paper the relations are stated primarily in terms of functors and their arguments, and represented by means of graphs. These graphs are not the familiar derivational trees with nodes consisting of *S*, *NP*, *VP*, etc.; instead, the nodes are functors and arguments – actual phrases rather than symbols from the metalanguage.

3.2. The contrast between the use of strings and graphs in a language is clearly illustrated by the representation of molecular structure in chemistry. Formulas for methane and benzene may be written in string form as CH_4 and C_6H_6 respectively. For a more complete description of the relations between the atoms within these molecules we may either make additional statements to accompany the string formulas or use structural formulas. The structural formulas for methane and benzene may be represented by the graphs (1) and (2):

(1) 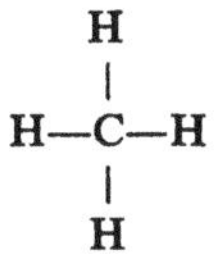 (2) 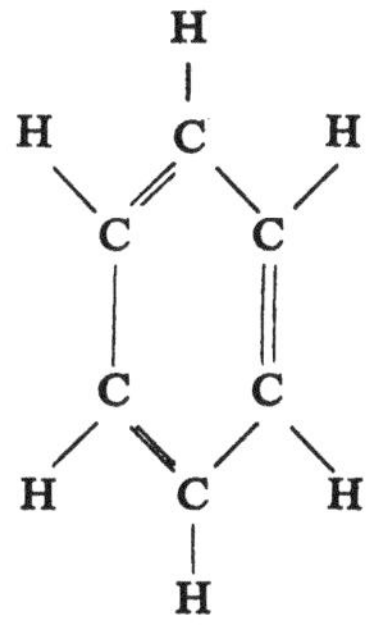

Methods for 'linearizing' such graphs have been worked out[2] so that one may start from a given node and trace a path through the graph to obtain a corresponding string of atomic symbols. Of course, the path which is traced does not reflect any inherent linearity in the structure of the molecule; neither does it reflect any particular order in the way the atoms were brought together to form the molecule. Thus the choice of a path through the graph in order to obtain a string representation is to a certain extent arbitrary; there may be many strings corresponding to a given graph.

In tracing a path through (1) or (2) it is necessary to return to certain nodes several times. Every return to a previously counted node has to be recorded at the appropriate place in the string, hence bookkeeping symbols are introduced to refer back to earlier positions in the string. (These reference symbols, unlike pronouns in natural languages, are not part of the language under investigation, but belong to the metalanguage.)

Certain information which is presented in an intuitively appealing manner by graphs is either lost altogether in a simple string formula (e.g. $C_6 H_6$ as compared with (2)) or is retrieved at the expense of augmenting the alphabet with bookkeeping symbols. The graph gives a direct representation of various relations between atoms within the molecule, and these relations constitute the structure. Essentially non-linear (i.e. non-stringlike) data can be put into string form with the help of suitable artifacts, but the

[2] See, e.g. Hiż, "A Linearization of Chemical Graphs".

multidimensional aspect of the structure of that data may be more readily observed by means of graph representation.

3.3. The use of graphs in the study of natural languages is not new. Even traditional high school grammars of the pre-transformational era used them in an informal way to 'diagram' sentences. E.g. the sentence *A man from Texas played his guitar for us* might be diagrammed as in (3):

(3)

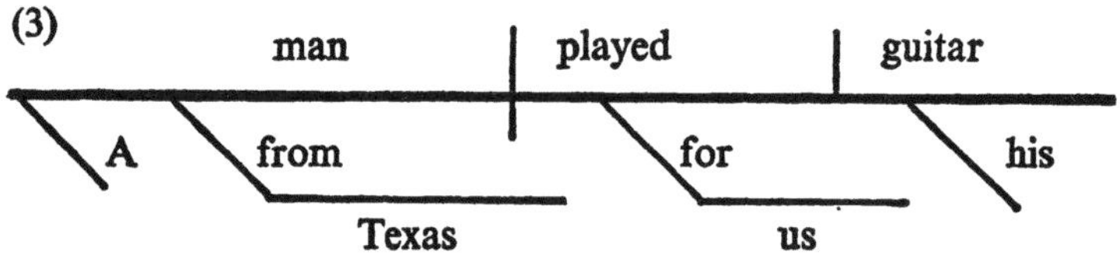

The vertical line which crosses the main horizontal line separates subject from predicate; the vertical line which terminates at the horizontal line separates direct object from verb; oblique lines indicate modifiers. The formal translation of diagrams such as (3) into graph-theoretical terms without losing the information conveyed by orientation of lines (horizontal, vertical, oblique), thickness of lines (dotted lines may also be used for certain types of modification) and length of lines, would be no simple task. Let us merely note that diagramming of this kind is an attempt to represent grammatical relations within a sentence directly in terms of the words that make up the sentence.

3.4. Traditional sentence diagramming could hardly be considered as the use of graphs in a technical sense. By comparison, in an article titled *Graphs for Linguistics*[3] Nelson Goodman speculated on possible applications of graph theory to linguistics. In particular, he considered the relation of MODIFICATION among words in a phrase: two words are connected by an arc in the graph provided either of them modifies the other. As an illustration he graphs this relation for the phrase *happy and healthy boys and girls*:

[3] In *Structure of Language and Its Mathematical Aspects*, ed. by R. Jakobson, 51-55.

(4)

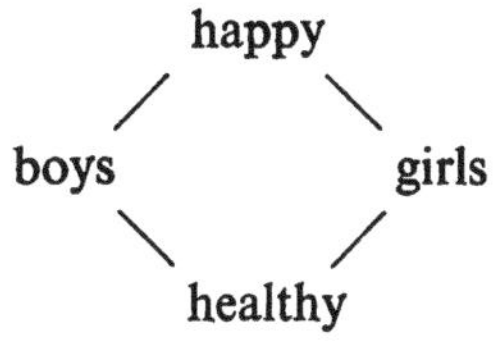

Regarding the more complicated example (5),

(5)

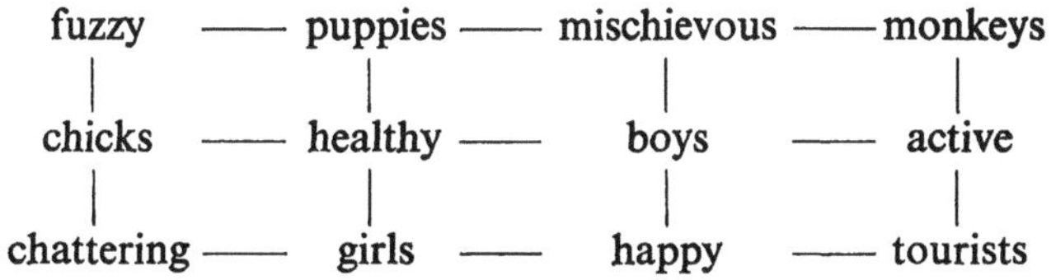

Goodman comments:[4] "This perfectly reasonable pattern might obtain, but it can hardly be embodied in an English sentence (except by taking repetitions of a word as identical). We could however construct a language having no such limitation."

These examples indicate some of the possibilities for the use of graphs in linguistics aside from the representation of derivations in a generative grammar.

3.5. In 3.1 arguments were presented against the characterization of sentences in a natural language as strings. However, as a starting point in the application of graphs let us first consider how graphs may be used for the direct representation of STRUCTURED STRINGS. The concept of a structured string has been described in detail by H. Hiż in *Grammar Logicism*[5] using functor-argument notation. Assignment of structure to a string x includes (i) the grouping imposed on x, (ii) the assignment of grammatical categories to certain substrings of x, and (iii) for each substring of x which belongs to a functor category, the identification of the arguments of that functor in x.

[4] Goodman, "Graphs for Linguistics", 54.
[5] In *The Monist* 51, 110-127.

((iii) is accomplished by writing numerals below the category symbols as we did in section 2.5, example (18).)

Consider the formula a—(b+c) = (a—b)—c from arithmetic. A grammatical analysis of this formula is given by the structured string:

(6)

a	—	b	+	c	=	a	—	b	—	c
N	(N;N_N)	N	(N;N_N)	N	(S;N_N)	N	(N;N_N)	N	(N;N_N)	N
4	5 4 3	1	3 1 2	2	11 5 10	6	8 6 7	7	10 8 9	9

The numerical variables a, b, c are of category N and occur only as arguments; +, — and = belong to functor categories. Note that if the third row is omitted in (6) there is no way to decide whether the first five symbols in the top row should be grouped (a—(b+c)) or ((a—b)+c): placing 1 under b informs us that b is an argument of + and not of —, since there is an N in the category symbol for + which is also tagged by a 1. It is not important what kind of tags are used in the third row, only that the proper matching is indicated. (This cross referencing will prove very useful later for the study of referentials in a natural language.)

The tags in the third row of (6) can be used to construct a graph which represents the structured string: If there is an occurrence of a symbol (X;Y_Z) in the second row properly flanked by occurrences of Y and Z, then replace the tags in the third row by arrows from X to the 'flanking' occurrences of Y and Z. E.g.:

(7)

b	+	c		
N	(N; N_N)	N	becomes	
1	3 1 2	2		

In this way there will be an arrow from an occurrence of a functor to each argument which satisfies that functor. Thus in (6) *b*+*c* is of category N and is an argument of the first occurrence of —, so that:

(8)

a — b + c

N (N;N_N) N (N;N_N) N becomes

4 5 4 3 1 3 1 2 2

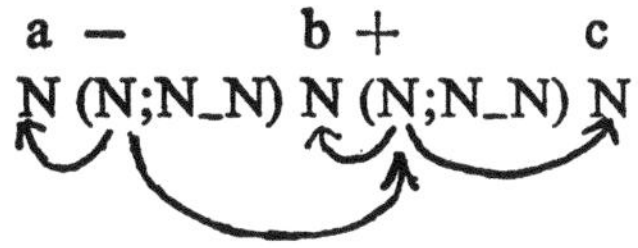

Following the same procedure throughout, (6) is replaced by:

(9)

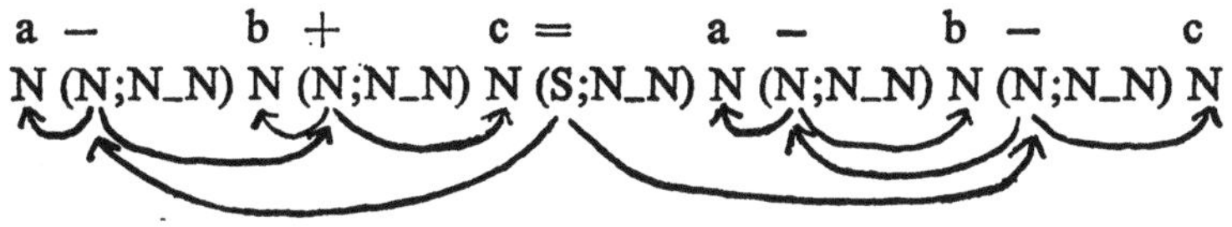

A graph is formed from (9) by taking each symbol in the first row as a vertex of the graph and taking a pair of vertices (x,v) as an arc if an arrow points from x to v.

Omitting all category symbols from (9) and rearranging the diagram, (10) is obtained:

(10)

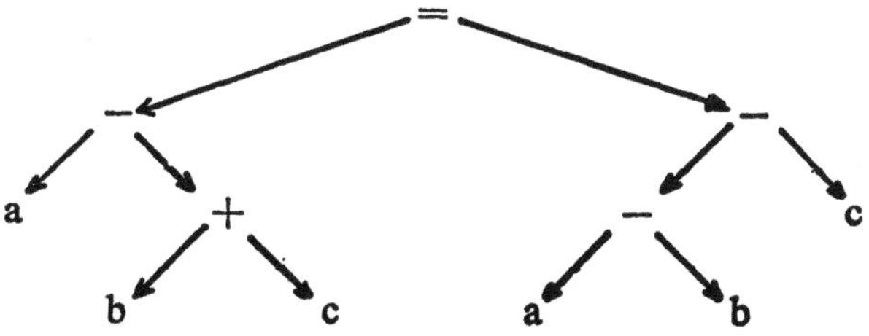

(Of course, for a complete description the category symbols must be included along with the nodes in (10).)

3.6. In this section a different approach will be used in the construction of a directed graph to represent (6). Let the statement:

(11) x is a phrase of grammatical category A

be abbreviated:

(11′) $A \Rightarrow x$

Let the somewhat lengthy statement:

(12) f is a functor which forms a phrase of grammatical category A out of arguments whose categories are B, C, ..., D

be abbreviated:

(12′)

$$A \Rightarrow f \boxed{\begin{matrix} B \\ C \\ \vdots \\ D \end{matrix}}$$

If b, c, ..., d are phrases of categories B, C, ..., D respectively, then by (11′) we may write:

(13) $B \Rightarrow b \qquad C \Rightarrow c \qquad \ldots \qquad D \Rightarrow d$

From (12′) and (13) a graph may be constructed by (i) replacing the category symbols B, C, ..., D in (12′) by the phrases b, c, ..., d respectively, and (ii) taking (f,b), (f,c), ..., (f,d) as the arcs of the graph:

(14)

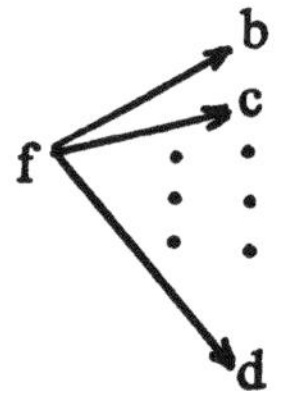

We may say that the graph (14) is DERIVED by the rules (12′) and (13). According to the full statement (12), the phrase which (14) represents is of category A and f is of category (A;BC ... D). (If the position of f relative to b, c, ..., d is given, this is indicated by '_' at the appropriate place in the category symbol and also in the 'box' in (12′).)

Now suppose that b is a functor of category $(B;B_1 \ldots B_\beta)$, c is a functor of category $(C;C_1 \ldots C_\gamma)$, ..., and d is a functor of category $(D;D_1 \ldots D_\delta)$. Then (13) is replaced by:

(15)

$$B \Rightarrow b\ \boxed{\begin{array}{c} B_1 \\ \vdots \\ B_\beta \end{array}} \qquad C \Rightarrow c\ \boxed{\begin{array}{c} C_1 \\ \vdots \\ C_\gamma \end{array}} \qquad D \Rightarrow d\ \boxed{\begin{array}{c} D_1 \\ \vdots \\ D_\delta \end{array}}$$

Replacement of B, C, ..., D in (12′) by the expressions on the right of the arrows in (15) still leaves us with category symbols inside the boxes. All category symbols must eventually be replaced by phrases which belong to those categories in order to obtain a graph on the set of phrases, but we may mark a step in the derivation of the graph as in (16):

(16)

$$f\ \boxed{\begin{array}{c} b\ \boxed{\begin{array}{c} B_1 \\ \vdots \\ B_\beta \end{array}} \\ \\ c\ \boxed{\begin{array}{c} C_1 \\ \vdots \\ C_\gamma \end{array}} \\ \vdots \\ d\ \boxed{\begin{array}{c} D_1 \\ \vdots \\ D_\delta \end{array}} \end{array}}$$

In order to replace ALL category symbols with phrases, 'terminal' rules of the form (11′) must be added to (12′) and (15). The final step in constructing the graph is to eliminate the boxes, forming a set of arcs as each box is eliminated. This may be done as follows: Starting with the outermost box, connect the functor on its left to each phrase inside which is not in any other box and then eliminate the box. (If there is no functor on the left of the box, simply eliminate the box.) Continue this procedure with each of the remaining boxes until all have been eliminated. The direction of each arc is from the functor on the left of the box to the phrase with which that functor is connected inside the box.

An example of a set of rules, (17i-vi), and a derivation of a graph from those rules is given below.

(17)

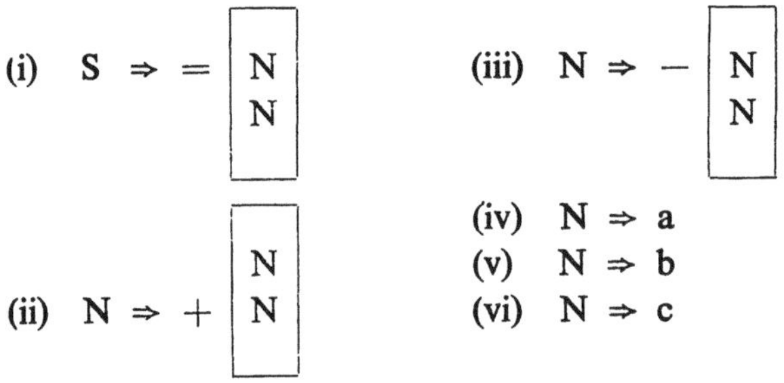

DERIVATION: From rules (17-i,iii) we get (18), and then by applications of (17-ii,iii,iv,vi), (19) is obtained:

(18) =

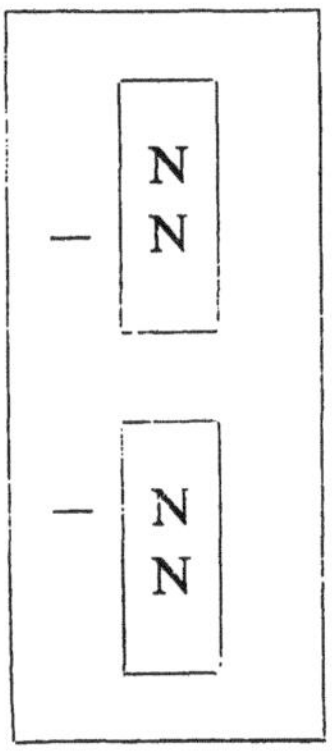

(19) =

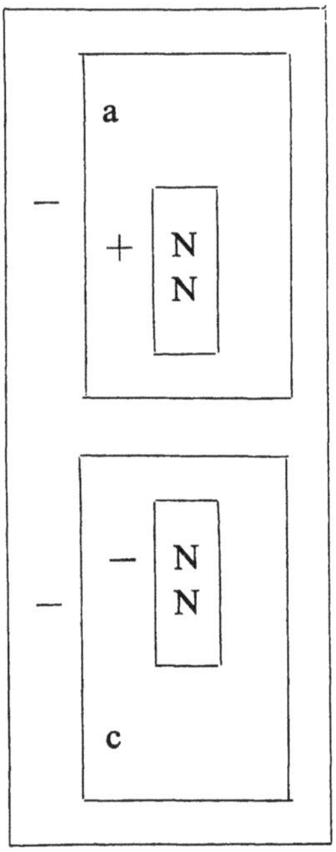

Finally, eliminate all category symbols by applying (14-iv,v,vi) to (19). This yields (20).

(20)

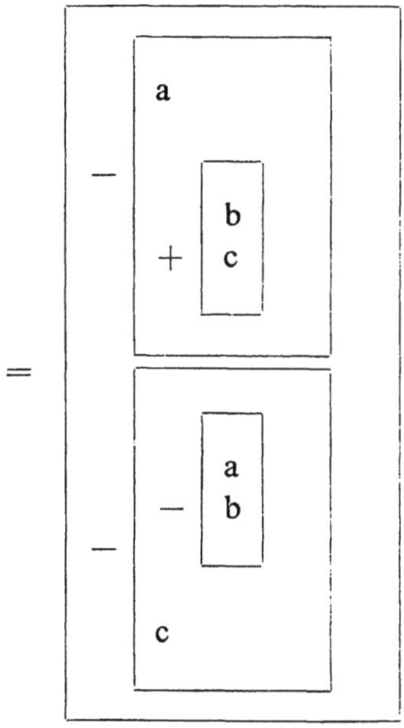

The resulting graph is therefore:

(21)

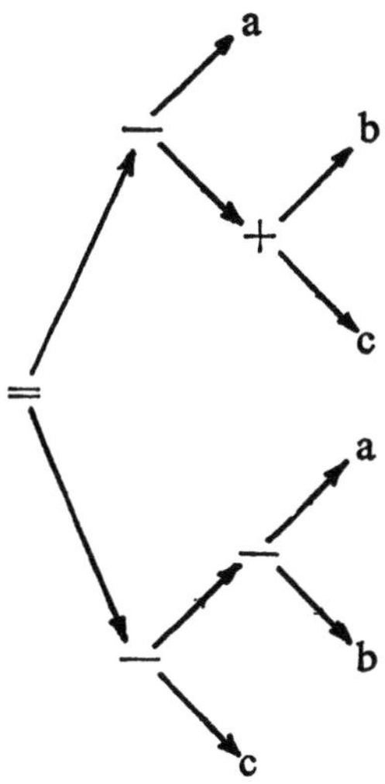

which is the same as (10). Thus identical graphs have been constructed by two different methods. In the construction of (10) we were given a structured string to start with and then translated the structured string into graph notation. In the case of (21) the graph was GENERATED from a set of rules. (Of course, there is a sense in which even the second method starts with the structured string – though not formally; i.e., the rules do not fall from the sky, but are set up with an eye to the end product.)

The graph ((21) or (10)) is a tree and the subtrees in it correspond to the grouping indicated by parentheses in (22):

(22) $$((a-(b+c)) = ((a-b)-c))$$

The generation of graphs by rules of the form (11′) and (12′) yields trees (and, possibly, isolated points). But THE METHOD USED TO CONSTRUCT (10) MAY RESULT IN GRAPHS OTHER THAN TREES WHEN APPLIED TO SENTENCES IN A NATURAL LANGUAGE. This will be shown in succeeding chapters.

There is an important difference between (21) and the derivation graphs commonly used in generative grammars. Each vertex in (21) is one of the symbols in the language under study, whereas vertices in the derivation graphs employed by generative grammarians consist of various abstract elements as well as actual phrases in the language being generated. In short, (21) is not a derivational tree, but a TERMINAL TREE.

Each non pendant vertex x in (21) 'dominates' a SUBTREE which represents a phrase y of some category A. E.g. + dominates the subtree:

(23)

```
  +
 ↙ ↘
b   c
```

which represents the phrase b + c of category N. The application of a non-terminal rule from (17) leads to the introduction of a subtree into the final graph; if this rule is $A \Rightarrow f\ \boxed{\vdots}$ then the subtree represents a phrase of category A, and f dominates the subtree. We could mark the category corresponding to the subtree by copying A along with f. Since each subtree is dominated by one and only one vertex (i.e. each subtree has exactly one root) the marking is unique. The use of the term *dominate* in this study is explained in 4.6.

In either (21) or (10) the arcs with initial points at the same vertex v are ordered corresponding to the order of the arguments of v. In a rule of the form (12′) the order of arguments of f is indicated by the order of the symbols inside the box, reading from top to bottom; the box with its column of symbols is an ordered n-tuple.

3.7. PARAMETRIC FORMS

A functor may have one or more parameters. The functor *equals x*, which may be written:

(24) $$= \langle x \rangle$$

has the single parameter x. Corresponding to (12′) we may write:

(25) $$S \Rightarrow\ = \langle x \rangle \boxed{N}$$

indicating that $= \langle x \rangle$ is a functor (with parameter x) which forms a phrase of category S out of one argument of category N. To construct a graph from (24), $= \langle x \rangle$ is taken as a single vertex. The graphs for the parametric form $= \langle x \rangle\ y$ and the usual $x = y$ are different, but the relation between them is given by means of the following definition. For all x,

(26) $$S \Rightarrow\ = \langle x \rangle \boxed{N} \quad \text{if and only if} \quad S \Rightarrow\ = \boxed{\begin{matrix} N_x \\ N \end{matrix}}$$

where the subscript x on N indicates that only the argument x is to be considered as a replacement for the N so marked. The left side of (26) yields (27), whereas the right side yields (28):

(27) = ⟨x⟩
↓
α

(28) =
↙ ↘
x α

where α is any phrase of category N, possibly x itself. We may say that (27) is a *transform* of (28) via the definition (26).

Since = ⟨x⟩ is of category (S;_N) and x is of category N, it follows that = ⟨x⟩ has the structure:

(29) = ⟨x⟩
((S;_N);_N) N
1 1

Thus (27) is a global representation; the vertex = ⟨x⟩ is complex and is represented by a graph: = → x.

A situation similar to (27) and (28) arises in English as shown in (30-31′):

(30) Fred is happy
N (S;N_A) A

(31) Fred is happy
N ((S;N_);_A) A
(S;N_)

(30′) is
↙ ↘
Fred happy

(31′) [is happy]
↘
Fred

The graph for *is happy* is:

(31″) is → happy

The graphs (31′) and (31″) are actually subgraphs (in fact, components) of a single graph corresponding to (31), which may be represented as:

(31‴) Fred ← [is → happy]

In terms of the definition of *graph* given in the appendix, the set X in this case is:

{Fred, is happy, is, happy}

and the graph, (X,F), is:

{(is happy, Fred), (is, happy)}

The relation between structures such as (30) and (31) will be discussed in detail in chapter 4, and graphs of the form (31‴) will be discussed further in 4.4.

A functor may have several parameters. Consider the three argument functor *between*. Let *x is between y and z* be abbreviated by b(x,y,z). The parametric form *between y and z* may be introduced by a definition similar to (26). For all y and z,

(32) $S \Rightarrow b\langle y,z\rangle \begin{array}{|c|}\hline N \\ \hline\end{array}$ if and only if $S \Rightarrow b \begin{array}{|c|}\hline N \\ N_y \\ N_z \\ \hline\end{array}$

where y and z are the parameters. If we add the rule N ⇒ x, the resulting graphs are:

(33) $\begin{array}{c} b\,\langle y, z\rangle \\ \downarrow \\ x \end{array}$ and (34) $\begin{array}{ccc} & b & \\ \swarrow & \downarrow & \searrow \\ x & y & z \end{array}$

3.8. OMISSION OF ARGUMENTS

The set of rules (35) may be used to generate graphs for sentences in propositional logic.

(35) $S \Rightarrow\ \supset \begin{array}{|c|}\hline S \\ S \\ \hline\end{array}$ $S \Rightarrow p$

$S \Rightarrow\ \equiv \begin{array}{|c|}\hline S \\ S \\ \hline\end{array}$ $S \Rightarrow q$

$S \Rightarrow\ \vee \begin{array}{|c|}\hline S \\ S \\ \hline\end{array}$ $\vdots$

$S \Rightarrow\ \sim \begin{array}{|c|}\hline S \\ \hline\end{array}$ $S \Rightarrow r$

The graph corresponding to:

(36) $\supset\ \equiv\ pq \supset pq$

is given in (37) – omitting the derivation:

(37)

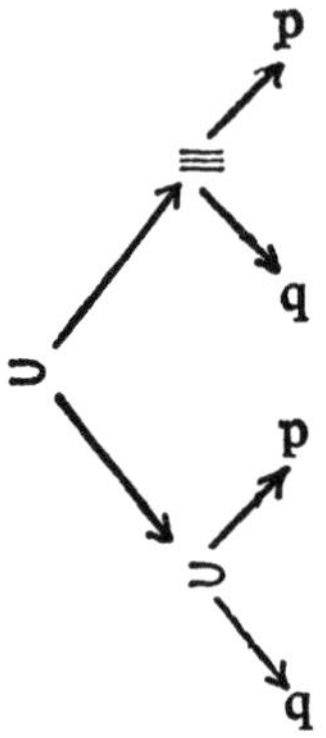

This shows a relation between $\equiv pq$ and $\supset pq$. There is also a relation, implicit in (36), between the functors $\equiv$ and $\supset$, namely, whenever $\equiv$ holds between two arguments so does $\supset$. This suggests writing $\supset \{\equiv\ \supset\}$. As a matter of fact, from the definition (38):

(38) $(\forall\varphi)(\forall\psi)(\supset \{\varphi\psi\}$ if and only if $(\forall p)(\forall q)(\supset \varphi pq\ \psi pq))$

substituting $\equiv$ for φ and $\supset$ for ψ, both of category (S;_SS):

(39) $\supset \{\equiv\ \supset\}$ if and only if $(\forall p)(\forall q)(\supset\ \equiv pq \supset pq)$

The expression $\supset \{\equiv\ \supset\}$ is of category S and the occurrences of $\equiv$ and $\supset$ inside the braces are both of category (S;_SS), therefore the left side of (39) has the structure (40):

(40)
$$\begin{array}{ll} \supset & \{\equiv \qquad \supset \quad \} \\ (S;\_(S;\_SS)(S;\_SS)) & (S;\_SS)(S;\_SS) \\ 3 \quad 2 \qquad\qquad 1 & 2 \qquad\qquad 1 \end{array}$$

Thus definitions such as (38) make it possible to express relations between functors directly without mentioning the arguments of those functors. The graph corresponding to (40) is:

(41)

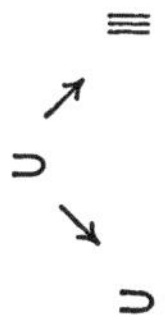

We may say that (41) IS A TRANSFORM OF (37) BY (39). This is an example of a zeroing transformation in which not just one, but four arguments have been zeroed. ALL pendant vertices in (37) were deleted in order to form (41).

4

STRUCTURED TEXTS

4.1. On what basis is structure assigned to a text in a natural language? Is the assignment to any extent arbitrary or, given a particular 'reading' of a text, is there a unique structure corresponding to that reading? The usual assumption has been that for different readings of a sentence different structures are assigned, and that no more than one structure is assigned for a given reading of a sentence. This one to one correspondence is supposedly obtained in a generative grammar by defining the structure in terms of a derivation of the sentence. The device for producing these derivations in a language L amounts to a recursive enumeration of the set of sentences (or readings of sentences) in L. But if L is a recursively enumerable set, it is the range of a recursive function. Presumably, grammarians should eventually be able to fashion a 'mechanical procedure' for turning out exactly the sentences of L – or many procedures since a recursively enumerable set can be enumerated by more than one set of rules. Chomsky foresaw the possibility of more than one grammar adequate for a given language and in *Syntactic Structures* he discussed the question of choosing a grammar. He rejected the notion of a linguistic theory providing a discovery procedure or a decision procedure for choosing THE correct grammar for a corpus and settled for an evaluation procedure: given a corpus and two grammars, the theory should provide criteria for choosing one of the two grammars over the other. However, once a grammar is selected it is supposed to provide one and only one derivation for each reading of a sentence.

One may question the assumption that the sentences of a natural language constitute a recursively enumerable set. Hockett has presented arguments[1] to refute this assumption, claiming on the contrary that NATURAL LANGUAGES ARE NOT WELL DEFINED and that they only possess certain 'stabilities'. One may also question the assumption that for an unambiguous sentence there is only one assignment of structure which is relevant within a particular grammar. The opposite point of view will be aired in 4.2. Finally, it is questionable whether a grammar of sentences alone is adequate in any event. The CONTEXT in which a sentence occurs affects our interpretation; a REFERENTIAL in one sentence may have a referend in another sentence; and CONTRASTIVE STRESS may relate phrases in different sentences. Texts[2] consisting of more than one sentence will therefore be analyzed and a grammatical unit between sentence and discourse will be defined.

4.2. The term STRUCTURED TEXT is used here to refer to the result of a particular assignment of functors and arguments to the elements of the text. Corresponding to a text T there may be various structured texts T_i, and each of these T_i is represented by a directed graph as described in chapter 3. The graphs are constructed after the functor-argument relations have been determined, following the method used to construct (10) from (6) in section 3.5.

Before discussing general criteria for deciding which phrases in a text are taken as functors and which phrases are taken as arguments of those functors, it will be helpful to examine a simple case. Consider the assignment of structure to the unambiguous text *John loves Mary*. Although there is no ambiguity in the usual sense, this text may be regarded from different points of view (these 'points of view' will be referred to here as INTERPRETATIONS). Three interpretations of *John loves Mary* are given below along with the functor-argument assignments that reflect those interpretations.

[1] Hockett, *State of the Art*.

[2] Richard Smaby has taken texts rather than sentences as the object of analysis in his *Paraphrase Grammars*.

(1)	John	loves	Mary	(*loves* as a relation
	N	(S;N_N)	N	between *John* and *Mary*)
	1	3 1 2	1	
(2)	[John	loves]	Mary	(*John loves* predicated
	N	((S;_N);N_)	N	about *Mary*)
	1	3 2 1	2	
(3)	John	[loves	Mary]	(*loves Mary* predicated
	N	((S;N_);_N)	N	about *John*)
	1	3 1 2	2	

(The brackets in (2) and (3) enclose complex functors of categories (S;_N) and (S;N_) respectively.)

It makes no more sense to ask which of the above is THE appropriate structure than it does to ask which interpretation is the 'correct' one (see the discussion in 3.7). Note also that (1) – (3) reflect answers to different questions: What is the relation of *John* to *Mary*? John loves who(m)? Who loves Mary? Let us accept each of these structures as relevant in the light of some interpretation of the sentence. In subsequent chapters more evidence will be presented to support this 'multistructure' point of view.

Since the assignment of more than one structure is usually taken to indicate an ambiguous text, it is necessary to clarify our use of the term *ambiguity* in this paper. The relation between (1), (2) and (3) may be viewed in the same way as that which holds between (4), (5) and (6) (only global structure is indicated in (5) and (6)):

(4) x	=	y	(5)	[x =]	y	(6)	x	[= y]
N	(S;N_N)	N		(S;_N)	N		N	(S;N_)

The parametric forms (5) and (6) are equivalent to (4), hence there is no semantic ambiguity. Likewise, no semantic ambiguity is indicated in (1) – (3). If a text is n-ways ambiguous semantically it is said to have n distinct READINGS; native speakers should be able to detect n different meanings for the text as a whole. The term *reading* will be retained in this sense and there may be many interpretations corresponding to a single reading. The resulting structural ambiguity (as in (1) – (3)) which does not reflect a semantic ambiguity will be referred to as PURE STRUCTURAL AMBIGUITY.

Pure structural ambiguity is not limited to declarative sentences, but is also applicable to interrogatives, imperatives, etc. Nor does it always involve a main verb. We shall see examples of pure structural ambiguity involving functors of various types.

4.3. Functors are used to describe the way phrases in a language combine to form other phrases. A grammar of a language should describe the RELATIONS that exist between the phrases which are combined to form coherent texts. These relations may be given a direct representation by means of functors. In order to establish criteria for deciding which phrases in a text are to be taken as functors, let us examine some typical relations.

If a phrase x is regarded as a predicate in a text T and phrases $y_1, \ldots, y_n$ as the arguments of x in T, then x may be taken as a functor and $y_1, \ldots, y_n$ as its arguments. Thus *snores* is normally taken as the predicate phrase in *Felix snores* and *Felix* as the argument; hence *snores* may be regarded as a functor whose argument is *Felix*. Graphically,

(7) Felix ← snores

Time honored though it may be, division of a sentence into subject and predicate is not without its pitfalls. Jespersen, discussing the 'topic-comment' definition of subject and predicate, gives the example of *John promised Mary a gold ring.*[3] What is the topic (subject) in this case? He comments: "... there are four things of which something is said, and which might therefore all of them be said to be 'subjects', namely (1) John, (2) a promise, (3) Mary, and (4) a ring". Alternatives (1), (3) and (4) are clearly suggested by the questions: Who promised Mary a ring? Who did John promise a ring? What did John promise Mary? (Note

[3] Jespersen, *The Philosophy of Grammar*, 146.

that if *Mary* is the subject then a DISCONTINUOUS predicate phrase results: *John promised ... a gold ring.*) The alternative (2) involves nominalization of the verb *promise*; but aside from this nominalization, the other alternatives suggest pure structural ambiguity as illustrated below:

(8) John promised Mary a gold ring
N (S;N_)

(9) John promised Mary a gold ring
(S;_N_) N

(10) John promised Mary a gold ring
(S;_N) N

(8) – (10) represent different interpretations of a single reading; another interpretation is represented by (11).

Certain words are clearly relational in nature. *Transitive verbs* (make, hit, describe, love) establish a relation between noun phrases. This was indicated, e.g., when *loves* was taken as a two argument functor in (1), section 4.2. The occurrence of *promised* in *John promised Mary a gold ring* establishes a relation between *John, Mary* and *a gold ring.* This may be indicated by assigning *promised* to the functor category (S;N_NN):

(11) John promised Mary (a gold ring)
N (S;N_NN) N N

Prepositions (on, between, of, from, etc.) are also relational words. In *the girl on first base*, *on* relates *the girl* and *first base.* A possible analysis of this noun phrase is obtained by taking *on* as a functor of category (N;N_N). *between* (or *between ... and*) may be considered a three argument functor.

Various types of MODIFIERS (or adjuncts) may be treated in the same way as predicates, although they may not be sentence forming. The modifier is taken as a functor and the modified phrase as its argument:

(12) (i) slept well [where V_i abbreviates (S;N_)]
V_i $(V_i;V_i\text{–})$
(ii) dank tarn
(N;_N) N
(iii) joy supreme
N (N;N_)
(iv) tomorrow the truth will out
(S;_S) ⏟ S

In the study of logic A FUNCTOR MAY BE REGARDED AS A CLASS: fx may be considered either as a phrase formed by the functor f together with its argument x, or as a statement to the effect that *x belongs to the class f*. In a natural language some occurrences of class nouns may be regarded as functors. The English sentence *John is a student* is expressed in Russian as Иван студент (John student). No Russian equivalent of *is a* is needed to make it clear that this is a statement that an individual belongs to a certain class. The class noun студент may be taken as a functor of category (S;N_). In the English equivalent the entire phrase *is a student* plays the same role, except that it is a complex functor:

(13) John [is a student] Иван студент
N (S;N_) N (S;N_)

Referentials are relational phrases. A pronoun, e.g. does not just 'take the place' of a noun; it may refer to one or more other phrases in the text (when not used deictically). A referential often establishes a connection between different sentences:

(14) The jobless millions were poor but proud. They had been chosen to save the country from inflation. It was their patriotic duty.

In (14), *they*, referring to *the jobless millions*, establishes a connection between the first two sentences: likewise, *it* connects the second and third sentences by referring to *to save the country from inflation*; the possessive pronoun *their* relates *the jobless millions* and *patriotic duty*, thus connecting the first and third sentences. The relation between phrases which is expressed by non-deictic use of a referential may be represented by taking the referential as a functor and the referend(s) as argument(s). (A phrase to which a referential refers is called a REFEREND; a non-linguistic object denoted by a referential is called a REFERENT.)

4.4. The discussion in 4.3 provides a guideline, but not a formal procedure, for making functor assignments in a text. Following these suggestions and comparing the given text with others which paraphrase it (and whose word composition is similar), some of the phrases in the text may be established as functors and other phrases as their arguments. The assignment of some phrases to certain categories is then used to determine the categories of the remaining phrases. This determination is not, however, unique. E.g. from the assignment given in (15):

(15)
$$\underbrace{\underbrace{\text{the girl}}_{\text{N}}\ \text{on first base}}_{\text{N}}$$

the phrase *on first base* may be considered as a functor of category (N;N_):

(16)
$$\underbrace{\text{the girl}}_{\text{N}}\ \underbrace{\text{on first base}}_{(\text{N};\text{N}\_)}$$

But *first base* is of category N, hence:

(17)
$$\underbrace{\text{on}\ \underbrace{\text{first base}}_{\text{N}}}_{(\text{N};\text{N}\_)}$$

and it appears that *on* is a functor that forms a phrase of category (N;N_) out of a phrase of category N:

(18)
$$\underbrace{\text{the girl}}_{\text{N}}\ [\underset{((\text{N};\text{N}\_);\_\text{N})}{\text{on}}\ \ \underbrace{\text{first base}}_{\text{N}}]$$

This corresponds to the interpretation of *on first base* as a MODIFIER of *the girl*. On the other hand, we may start from (19):

(19) the girl on first base

N N

N

and assign *on* to the category (N;N_N). This corresponds to the interpretation of *on* as a relational phrase in the manner discussed in the paragraph on prepositions, section 4.3. Both of these assignments of structure reflect valid interpretations – as well as the assignment of *on* to the category ((N;_N);N_). It is another example of pure structural ambiguity.

In chapter 1 it was pointed out that a functor may itself serve as argument of another functor. It is important to distinguish the use of a FUNCTOR as argument from the use of a FUNCTOR-ARGUMENT PHRASE as argument of another functor. The following example illustrates the difference. Consider the structure of *wide red car* (a red car which is wide) and that of *deep red car* (a car whose color is deep red):

(20) wide red car
(N;_N) (N;_N) N
3 2 2 1 1

(21) deep red car
((N;_N);_(N;_N)) (N;_N) N
3 2 1 1 2

The argument of *wide* is *red car*, but the argument of *deep* is just the functor *red*; hence the tag 1 is placed under '(' in the index for *red*, not under N. The corresponding graphs are:

(20′) wide ⟶ red ⟶ car

(21′) [deep ⟶ red] ⟶ car

(21′) is similar to (31‴) in 3.7; it is a graph with two components. The following example requires a graph with three components.

In this example *deep red car* is itself taken as a functor since it occurs as a modifier of another word.

(22)	deep	red	car	syndrome
	(((N;_N);_N);_(N;_N))	(N;_N)	N	N
	4 3 2 1	1	2	3

(22′) [[deep ⟶ red] ⟶ car] ⟶ syndrome

In terms of the definition of *graph* given in the appendix, the set X is:

{deep, red, car, syndrome, deep red, deep red car}

and the graph, (X,F), is:

{(deep, red), (deep red, car), (deep red car, syndrome)}

Complex functors give rise to 'complex vertices' in the graph representation (a COMPLEX FUNCTOR is one which has at least one other functor among its parts). Complex functors will be enclosed in square brackets where necessary to avoid confusion.

4.5. The assignment of structure to a text T reflects not only relations between phrases within T, but also the relation of T to other texts. When these 'external' relations are taken into account the multistructure point of view (section 4.2) receives additional support. If $\mathscr{R}$(T) is a reading of T, the structures T_i, T_j, ... corresponding to different interpretations of $\mathscr{R}$(T) may be appropriate for describing the relation of $\mathscr{R}$(T) to readings of various other texts. E.g. *John promised Mary a gold ring* bears a close relation to each of the following: (i) *It was John who promised Mary a gold ring*, (ii) *It was Mary that John promised a gold ring* and (iii) *It was a gold ring that John promised Mary.* The structures best suited to describe the relation of the given text to (i), (ii) and (iii) are (8), (9) and (10) respectively (section 4.4).

Relations between texts will be discussed in detail later. For now, let us merely note that intertextual relations are tied in with the ancient problem of assigning phrases to grammatical categories. (See also chapter 1, pages 13 and 14.)

4.6. Terminological note: In a given text, the phrase formed by a functor f together with its arguments a_1, ..., a_n will be called a FUNCTOR-ARGUMENT PHRASE. The functor will be said to DOMINATE the entire functor-argument phrase. For example, in (20) *wide* dominates *wide red car*; in (22) *deep* dominates *deep red*, *deep red* dominates *deep red car*, and *deep red car* dominates *deep red car syndrome*. The functor must be part of the phrase which it dominates; e.g., in (20) we will not say that *wide* dominates *red*, *car*, or *red car*. Thus the main functor of a phrase is said to dominate the phrase.

5

SHARING

5.1. REFERENTIALS

5.1.1 A grammar which treats referentials by deriving sentences containing referentials from other sentences which do not contain them may fail to characterize the relation of a referential to its referend(s) in the same text. Consider the text:

(1) John turned on the TV. He poured a beer. His eyelids drooped.

The second sentence may be derived from *John poured a beer* and the third from *John's eyelids drooped.* But a grammar which only generates sentences misses the relation which the referentials establish between the three sentences in (1). Additional apparatus is required since neither the second nor third sentence is derived from the first.

In this book the relation of a referential to its referend(s) in a text is given a direct representation by taking the referential as a functor and its referend(s) as argument(s). The resulting combination will be called a REFERENTIAL PHRASE. Thus in (1) *John ... he* is a referential phrase. Such phrases will usually be discontinuous, though not always. E.g. *The note was written by John himself* contains the referential phrase *John himself.* The left hand argument of *poured* in (1) is the entire referential phrase, not just the referential *he.* In this way we give syntactic recognition to something which is obvious semantically when the text rather than the isolated sentence is taken into consideration. We are, in effect, treating the second sentence as though it were *He (John) poured a beer.* But there is no need to insert the referend since it is already present in the text. Neither is there any need to derive the second sentence of (1) from some abstract object in order to explain the reference of *he*. We simply extend the notion of SYNTACTIC CONNECTEDNESS to texts. The semantic sensitivity of the grammar is thereby sharpened without positing any additional elements beyond the text itself.

Since *John ... he* in (1) is the first argument of *poured*, and *poured* is of category (S;N_N), the referential phrase is assigned to the category N; it plays the same role as *John* in *John poured a beer*. The referential *he* is therefore a functor which takes an argument of category N *(John)* and forms a phrase of category N *(John ... he)*; i.e. *he* is of category (N;N). The position indicator '_' is omitted from the category symbol. However, the tags under the category symbols identify the position of referend relative to referential. The following partial analysis of (1) may now be given:

(2)	John	turned on	the TV	He	poured	a beer
	N	(S;N_N)	N	(N;N)	(S;N_N)	N
	1	3 1 2	2	4 1	6 4 5	5
	1					

Graphically,

(3)

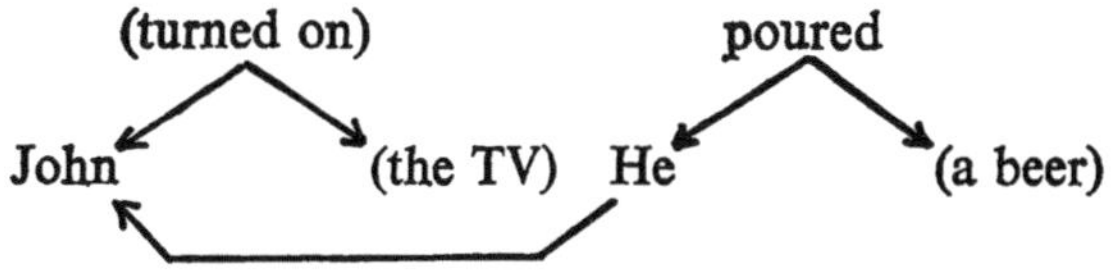

Note that the single occurrence of *John* serves as an argument of *turned on* as well as of *he*. Hence the tag 1 is written twice under *John* in (2). The use of one occurrence of a phrase as an argument of more than one functor will be referred to as ARGUMENT SHARING. In the graph (3) argument sharing is indicated by the fact that the vertex *John* is the terminal point of two arcs. In general,

(4) If an occurrence of a phrase serves as (exactly) one argument of each of n functors in a structured text T_i then in the graph of T_i the corresponding vertex will have an indegree of n.

There will be n tags under such an argument occurrence in T_i to indicate that this occurrence is involved in n cancellations. The converse of (4) does not hold (see e.g. (90), 5.2.2).

As for the third sentence of (1), it is tempting to assign the global structure (5):

(5) $\underbrace{\text{his eyelids}}_{\text{N}}$ drooped (S;N_)

which would result in:

(6)
his eyelids drooped
(N;N) N (S;N_)
8 7 7 9 8

But (6) gives no indication of the fact that the possessive pronoun *his* expresses a relation between two phrases, *John* and *eyelids* in the text (1). In this respect *his* is like the possessive suffix in *John's eyelids*, except that *his* is separated from *John* by other phrases. (Of course, they may be contiguous in other contexts; e.g. *give John his sedative.*) The possessive suffix 's together with a preceding phrase of category N and a following phrase of category N forms a phrase of category N:

(7)
John 's eyelids
N (N;N_N) N

Similarly, *his* relates *John* and *eyelids*:

(8)
John ... his eyelids
N (N;N,_N) N

(The comma to the left of the position indicator in the category symbol indicates that the functor is not contiguous with its first argument, but is separated from it by phrases on the left.) It follows that if the whole text is taken into account in assigning structure then instead of (6) the result is:

```
(9)          John ... his        eyelids drooped
              N     (N;N,_N)   N     (S;N_)
              1      8  1  7   7      9 8
```

Combining (9) and (2) we get:

```
(10) John turned on the TV He      poured    a beer
      N   (S;N_N)    N   (N;N) (S;N_N)    N
      1    3 1 2     2    4 1   6 4 5     5
      1
      1
     His        eyelids drooped
     (N;N,_N)    N     (S;N_)
      8 1 7      7      9 8
```

and the corresponding graph:

(11)

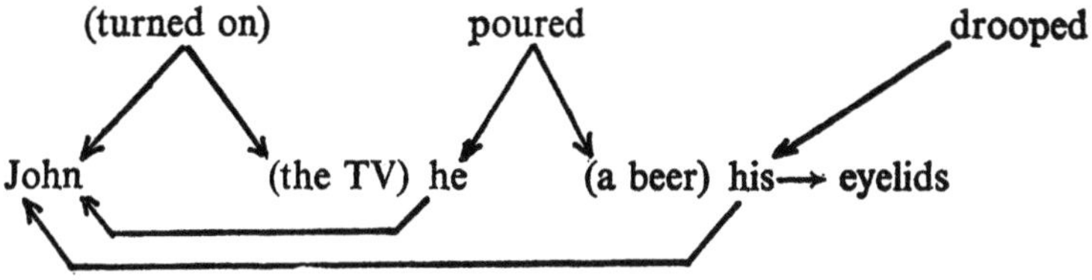

The preceding analysis entails the assignment of phrases such as *John ... he poured a beer* to the category S. The sharing of a phrase by more than one sentence means that we cannot say that the text (1) is a *sequence* of three sentences, only that it consists of three sentences. A more striking example of this interlacing of sentences is seen in the text (12):

(12) Elizabeth was standing opposite the antique mirror. Suddenly the Count entered the drawing room. She noticed that his reflection was missing.

Taking the referends into account, the third sentence of (12) may be paraphrased by:

(13) She (Elizabeth) noticed that his (the Count's) reflection was missing.

Therefore in (12) the discontinuous phrase:

(13′) Elizabeth ... the Count ... She noticed that his reflection was missing

is assigned to the category S. (13′) may be referred to as a REFERENTIALLY EXTENDED SENTENCE.

The symbol S is used in linguistics nowadays to label a variety of objects, presumably all related in one way or another to things that native speakers (who are not linguists) would be willing to call sentences. The use of the term SENTENCE will be discussed in 6.2. Meanwhile we simply note that phrases such as (13′) are included in the category S. Further examples of referentially extended phrases will be given in the sections which follow.

There is one other matter concerning the interpretation of referentials as functors. It may be desirable to assign *he* and *she* to different categories, say $(N_m;N_m)$ and $(N_f;N_f)$, where N_m and N_f are masculine and feminine subcategories of N. For example,

(14)	John ...	he	poured	a beer
	N_m	$(N_m;N_m)$	(S;N_N)	N
	1	4 1	6 4 5	5

Normally cancellation of indices takes place between identical indices, but in (14) the index N_m cancels with N. Of course, N_m is a subset of N, hence the requirement that the first argument of *poured* be of category N is met by any phrase belonging to N_m. This type of matching (or cancellation) of indices illustrates a general principle:

(15) In the representation of a structured text an occurrence of an index Z may cancel with Y_i in the index $(X;Y_1 \ldots Y_n)$ if $Z \subset Y_i$.

5.1.2 The sentence *John looked before he leaped* may be analyzed as in (16):

(16)

John	looked	before[1]	he	leaped
N	(S;N_)	(S;S_S)	(N;N)	(S;N_)
1	2 1	5 2 4	3 1	4 3
1				

The corresponding graph is:

(17)

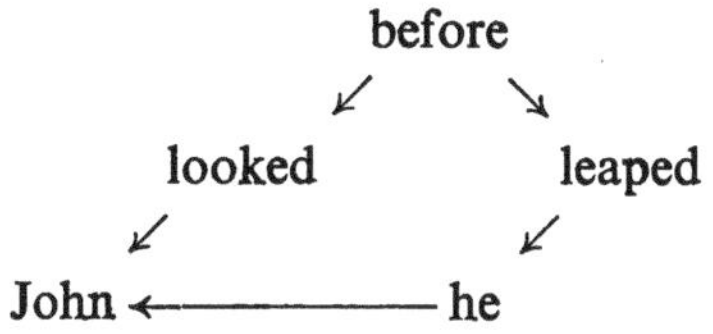

In (17) there are two distinct paths from *before* to *John*, therefore the graph contains a CYCLE – but no circuit. Argument sharing within a text may lead to cycles in the functor-argument graph, but not always. E.g., there are no cycles in (11) although there is argument sharing. Argument sharing is a necessary but not a sufficient condition for the presence of cycles in the corresponding graph. Other examples involving cycles are given below:

(18)

Phillip	washed	himself
N	(S; N_N)	(N; N)
1	3 1 2	2 1
1		

(18′)

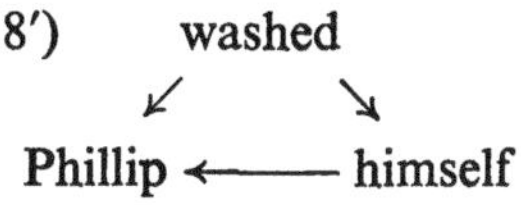

(19)

Mary	applied	her	makeup	while	John	waited	impatiently
N	(S;N_N)	(N;N,_N)	N	(S;S_S)	N	(S;N_)	(S;S_)
1	4 1 3	3 1 2	2	8 4 7	5	6 5	7 6
1							

[1] Strictly speaking, the functor *before* is not 'between' its argument phrases as the position indicator in the category symbol suggests since part of the second argument actually lies to the left of the functor. However, we shall use the position indicator in this way provided that the part of the argument contiguous with the functor on the right includes the main functor of that argument phrase (and similarly for the 'left hand' argument). E.g. the main functor of the right hand argument of *before* in (*16*) is *leaped* which is included in *he leaped*, the part of the argument contiguous with *before* on the right.

(19′)

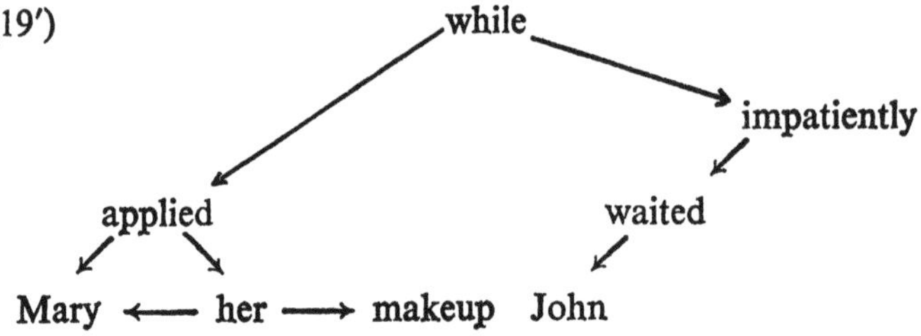

(20) John laughs at [his own] jokes

N (S;N_N) (N;N,_N) N

1 4 1 3 3 1 2 2

1

The word *own* in (20) combines with *his* to form a phrase of the same category as *his*. Abbreviating the symbol (N;N,_N) by α, the structure of *his own* is:

his own

α (α;α_)

and the final result, graphically, is:

(21)

5.1.3 *Relative pronouns*

5.1.3.1 First consider the role of relative pronouns in RESTRICTIVE relative clauses. In (22) each relative pronoun REFERS to the noun phrase preceding it.

(22) a. an actor who lives his part
b. a new book which I read
c. birds which we saw nesting in the tree
d. the roof that leaked
e. people who sing

Suppose the relative pronoun is regarded as a one argument functor whose argument is the noun phrase to which it refers; then it would fall in the same category as *he* or *she*. E.g. (22)e would be assigned the structure:

```
(23)            people who    sing
                  N   (N;N) (S;N_)
                  1     2 1   3 2
```

and the entire phrase, which functions as a noun phrase, would be of category S. On the other hand, if *who* is not regarded as a functor but is simply assigned to N, then the result is:

```
(24)            people who sing
                  N    N  (S;N_)
```

where *who sing* is of category S and *people* is left dangling. In neither case does the analysis of the relative pronoun lead to a satisfactory determination of the structure of the restrictive relative clause.

A relative pronoun in a restrictive clause does not just refer to a noun phrase; it relates the noun phrase to another phrase which qualifies or restricts the noun phrase. Compare (22)e with *people sing*: without the relative pronoun acting as a mediator, this amounts to a predication about all people. But the relative pronoun combines with *sing* to pick out a subset from the set of all people. Thus *who sing* functions like an attributive adjective. Jespersen, referring to *the rose which is red*, remarks "the function of the relative is precisely that of making the whole thing into an adjunct (an attribute, an epithet)".[2]

The mediating role of relative pronouns suggests treating them as two argument functors. In this respect they are like possessive pronouns. Each relative in (22) combines with the preceding noun phrase and the following predicate phrase to form a noun phrase:

[2] Jespersen, *The Philosophy of Grammar*, 114.

(25) the roof that leaked a book which I read
N ? (S;N) N ? (S;N)
N N

This determines the category of the relative pronoun as (N;N_(S;N)). The structure assigned to (22)e is therefore:

(26) people who sing
N (N;N_(S;N)) (S;N)
1 3 1 2 2

The foregoing analysis of relative pronouns in restrictive clauses may be represented schematically as in (27):

(27) x wh(x) y
N (N;N_(S;N)) (S;N)

or, graphically,

(28) wh(x)
↙ ↘
x y

Alternatively the relative pronoun may be assigned to the category ((N;N_);_(S;N)), another example of pure structural ambiguity:

(29) x wh(x) y
N ((N;N_);_(S;N)) (S;N)
1 3 1 2 2

Or, graphically,

(29′) x ← [wh(x) → y]

This shows that [wh(x) y] behaves like an adjective, forming a noun phrase out of a noun phrase.

Since the entire phrase in either (27) or (29) is of category N, it may serve as argument of a functor of category (S;N_):

(30) people who sing [are happy]
N (N;N_(S;N)) (S;N) (S;N_)
1 3 1 2 2 4 3

(30′) who ← —— [are happy]

↙ ↘

people sing

Or, alternatively,

(31) people ← [who ————→ sing] ← [are happy]

N ((N;N_);_(S;N)) (S;N) (S;N_)

1 3 1 2 2 4 3

Further examples of (27) are given in (32) – (37):

(32) a house which I painted

N (N;N_(S;N)) (S;N)

1 3 1 2 2

(33) the man who pushed the button

N (N;N_(S;N)) (S;N)

1 3 1 2 2

(34) flowers that [bloom early]

N (N;N_(S;N)) (S;N)

1 3 1 2 2

(35) snow which had been lying on the lawn all week

N (N;N_(S;N)) (S;N)

1 3 1 2 2

(36) the bird which distressed the man who

N (N;N_(S;N)) ((S;N);_N) N (N;N_(S;N))

6 7 6 5 5 4 3 4 3 2

wore (a black hat)

((S;N);_N) N

2 1 1

The graph corresponding to (36) is:

(36′) which

↙ ↘

(the bird) distressed ——→ who

↙ ↘

(the man) wore → (a black hat)

In chapter 2 we discussed the structure of *To a man who was poor John gave money*. The following analysis incorporates the treatment of restrictive relative clauses given above:

(37)

To	a man	who	was	poor	John	gave	money
to	N	(N;N_(S;N))	((S;N);_A)	A	N	(S;toNN_N)	N
1	2	5 2 4	4 3	3	6	8 1 5 6 7	7

(37′)

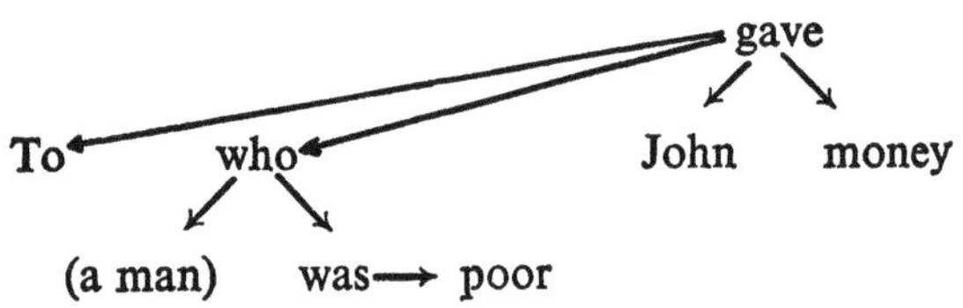

An entire relative clause may serve as a shared argument. Thus in *people who whistle amuse themselves* the referend of the reflexive pronoun is not *people* but *people who whistle*:

(38)

people	who	whistle	amuse	themselves
N	(N;N_(S;N))	(S;N)	(S;N_N)	(N;N)
1	3 1 2	2	5 3 4	4 3
	3			

(38′)

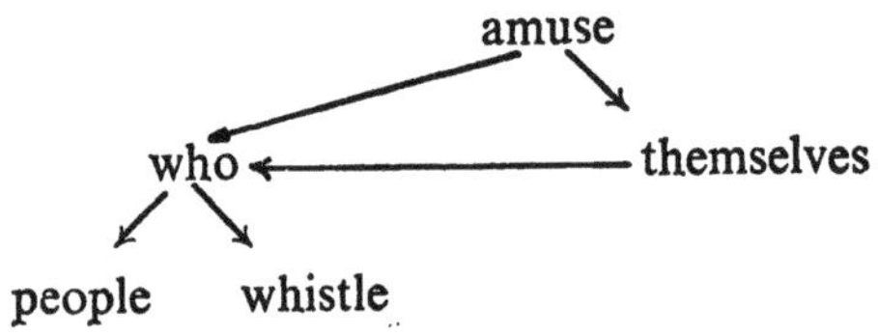

5.1.3.2 Nonrestrictive relative clauses

Nonrestrictive relative clauses are normally set off by pauses in speech and by commas in writing. For example, the relative clause in (39) is restrictive while that in (40) is nonrestrictive:

(39) Men who are fallible make mistakes
(40) Men, who are fallible, make mistakes

The role of intonation in differentiating sentences such as these will be discussed in 6.2.2. For the moment, let us consider only the role of the segmental elements. (40) may be paraphrased by:

(41) Men are fallible and men make mistakes

Just as *men are fallible* is a constituent of (41), so we may take the corresponding discontinuous phrase as a constituent in (40), letting *men* serve as argument of *are fallible*. This reflects the fact that semantically (40) is being used to make a predication about all men, unlike the situation in (39). A partial analysis of (40) is then:

(42)

men	who	[are fallible]
N	?	(S;N)
1		2 1

The question now is what category *who* should be assigned to. Consider another paraphrase of (40):

(43) Men (they are fallible) make mistakes

Here *they* plays exactly the same role as *who* in (40). This suggests taking *who* in the nonrestrictive clause as a one argument referential whose argument is *men*. The category assignment for *who* in nonrestrictive clauses is therefore different from that in restrictive clauses; the former need only show the relation of referential to referend while the latter must reflect the additional mediating role of the relative pronoun as explained in 5.1.3.1. We may now complete (42):

(42′)

men	who	[are fallible]
N	(N;N_)	(S;N_)
1	2 1	3 2

Since *make mistakes* is also predicated about *men* in (40), the full analysis is:

(44) men who [are fallible] [make mistakes]
N (N;N_) (S;N_) (S;N)
1 2 1 3 2 4 1
1

The general form corresponding to (44) is:

(45) x wh(x) y z
N (N;N_) (S;N_) (S;N)
1 2 1 3 2 4 1
1

(45′)

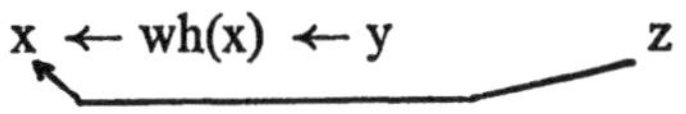

Compare this with the form involving restrictive clauses:

(46)

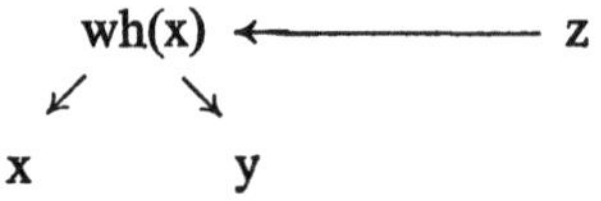

Now consider the role of *which* in the following examples:

(47) Birds eat insects which annoy me
(48) Birds eat insects, which annoy me
(49) Birds eat insects, which annoys me

According to the preceding analysis, *which* in (47) would be assigned to the category (N;N_(S;N)), and in (48) to the category (N;N_). That is, *which* occurs in a nonrestrictive clause in (48) and functions simply as a referential whose referend is *insects*. But (49) presents a new situation: *which* does not refer to *insects*, as is evident from the fact that the verb is in the third person singular. (49) may be paraphrased by

(50) Birds eat insects; this annoys me

or,

(51) Birds eat insects and birds' eating of insects annoys me

which in (49), like *this* in (50), is a referential whose referend is *birds eat insects*; in either case the proword has the same effect as the nominalization in (51). It follows that *which* in (49) is a functor which, together with a phrase of category S (the referend) forms a phrase of category N. The assignment of that occurrence of *which* to the category (N;S_) reflects its role as a nominalizer as well as a referential. Thus in (47), (48) and (49) *which* occurs in the categories (N;N_(S;N)), (N;N_) and (N;S_) respectively.

5.1.4 *Sameness of referent*

The treatment of pronouns as functors is not without precedent. As pointed out in chapter two, Lambek assigned *he* the type S/(N\S) and *him* the type (S/N)\S in order to distinguish pronoun in subject position from pronoun in object position. But these are our functor categories (S;_(S;N_)) and (S;(S;_N)_) respectively. Thus in:

(52) he yawned
S/(N\S) N\S

he is interpreted as a functor whose argument is the VERB *yawned*. This is quite different from the interpretation of *he* as a functor whose argument is the noun phrase to which it refers, but it does indicate a more flexible attitude toward functorship. In fact, Lambek's system included the EXPANSION RULES:

(53) (i) x → y/(x\y)
(ii) x → (y/x)\y

From (53-i) we may obtain:

(54) N → S/(N\S)

Thus in place of the usual:

(55) God exists (*exists* as operator on *God*)
N N\S

one may also write:

(56) God exists (*God* as operator on *exists*)
S/(N\S) N\S

From this point of view it would appear that there is nothing sacrosanct about the idea that some terms occur only as arguments, never as functors. (See H. Hiż's "On the Abstractness of Individuals"[3] for a discussion of this idea in logic.)

Compare the texts (57) and (58):

(57) Bill snores. Helen is annoyed with Bill.
(58) Bill snores. Helen is annoyed with him.

Assume that *him* in (58) refers to *Bill.* That is, *him* and *Bill* denote the same person (have the same REFERENT). If (58) is a paraphrase of (57), the second occurrence of *Bill* in (57) has the same referent as the first. We may equally well say that THE SECOND OCCURRENCE OF *Bill* IN (57) REFERS TO THE FIRST OCCURRENCE. This is tantamount to taking the second occurrence as a referential whose referend is the first occurrence. For example,

(59)

Bill	snored.	Helen	elbowed	Bill.
N	(S;N_)	N	(S;N_N)	(N;N)
1	2 1	3	5 3 4	4 1
1				

Thus the referential system may be expanded to include repetitions of a noun which have the same referent:

(60) If $x_1, x_2, \ldots, x_n$ are occurrences of the same noun in a text, all having the same referent, and x_1 is the first occurrence, then $x_2, \ldots, x_n$ may be taken as functors of category (N;N), and x_1 is the (shared) argument of these functors.

Both (60) and (53) have the effect of enabling us to treat nouns as functors even when they do not occur as modifiers, but these rules are quite different otherwise: they are different in form, they assign a phrase to different categories, their consequences differ (e.g. (60) leads to argument sharing), and they are motivated by different considerations.

[3] In *Identity and Individuation.*

5.1.5 In the examples given so far a referential plays two roles: (i) it refers to one or more phrases in the text, and (ii) it forms part of an argument of some other functor. Selection of a referential in a text depends to some extent on both of these roles. As an illustration of (ii), *he*, *she* and *they* normally occur in first argument position whereas *him*, *her* and *they* do not. (i) is reflected in number and gender agreement between referential and referend. Another important aspect of (i) is the SCOPE of the referential: the form of a referential may give a clue as to what part of a certain phrase is to serve as its referend. This may not pose any problem if the referend is an expression that serves as a proper name:

(61) John read *A Tale of Two Cities* and Mary read it too
(62)*John read *A Tale of Two Cities* and Mary read one too

Of course, (62) is not acceptable with the name of the book as referend of *one*. In the next example number agreement is the clue to the scope of the referential.

(63) John read books every evening and Mary read them too
(64) John read books every evening and Mary read one too

(64) is acceptable only if the scope of *one* is *book*, not *books*. Compare the referends in (65) and (66):

(65) John read this book and Mary read it too
(66) John read this book and Mary read one too

It is possible in (66) that Mary and John read the same book, but if that is the intended message then (65), not (66), would be used to convey it. The scope of *one* in (66) is therefore *book*, not *this book*. A similar situation arises when an indefinite article precedes the noun:

(67) John read a book and Mary read it too
(68) John read a book and Mary read one too[4]

[4] A transformational analysis of this example is given in Beverly Robbins' *The Definite Article in English Transformations*, section 4.4.

The structural difference resulting from the use of *one* rather than *it* in this example is illustrated below:

(67′)	a	book...	it
	(N;_N)	N	(N;N)
	2 1	1	3 2

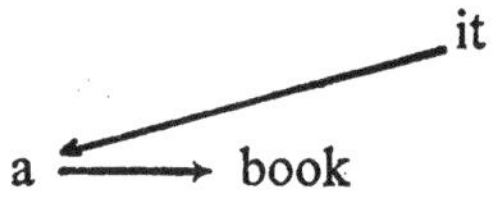

(68′)	a	book...	one
	(N;_N)	N	(N;N)
	2 1	1	4 1
		1	

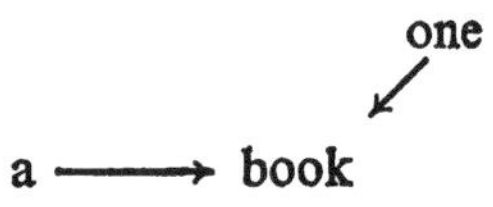

5.1.6 *Definite article*

In 5.1.3 we encountered noun phrases of the form *the x* in the context *the x wh(x) y*, where y helped to identify x. When *the x* does not occur in this context it most likely REFERS to another occurrence of x in the same text. In such a case *the x* plays the same role as *it*, *he*, etc. For example, (69) is a paraphrase of (70):

(69) Jim picked up a rock. It was very shiny.
(70) Jim picked up a rock. The rock was very shiny.

In (69) *it* refers to *a rock* and is in the first argument of *was*; the same is true of *the rock* in (70). Suppose *the rock* is regarded as a referential like *it*:

(71) Jim picked up	(a rock).	(The rock)	was	very shiny.
	N	(N;N)	(S;N_A)	A
	1	2 1	4 2 3	3

Consider the status of *the* according to this analysis. The traditional assignment of *the* to the category (N;_N) would miss the obvious relation which *the* established between the occurrences of *rock* in the first and second sentences of (71). THE FUNCTION OF *the* IS TO FORM A REFERENTIAL OUT OF A NOUN; it is a functor of category ((N;N);_N). Hence a more detailed analysis of (70) is:

(72)

Jim	picked up	a	rock	The	rock	was	very	shiny
N	(S;N_N)	(N;_N)	N	((N;N);_N)	N	(S;N_A)	(A;_A)	A
1	4 1 3	3 2	2	6 3 5	5	9 6 8	8 7	7
		3						

(72')

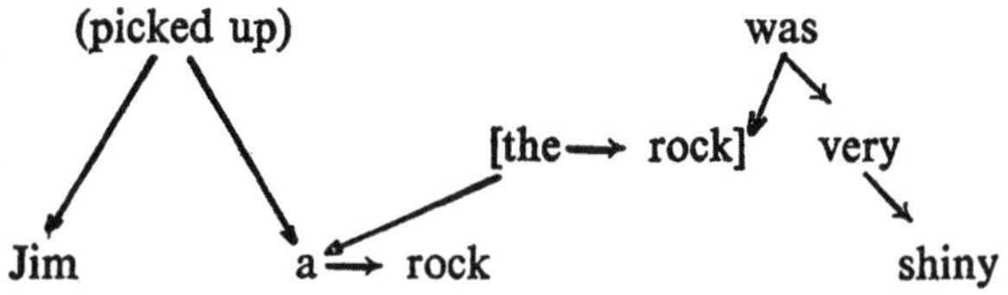

In summary, *the* together with *rock* forms a referential *the rock* whose referend is the phrase *a rock* in the preceding sentence; *a rock* is a shared argument since it also occurs as second argument of *picked up*.

Some occurrences of *this, that, these* and *those* may also be assigned to ((N;N);_N). E.g. in:

(73) Saucers dropped out of the sky into our yard and little people emerged from these objects

these together with *objects* forms a referential *these objects* whose referend is *saucers*. And in:

(74) Podunk is a great place to visit, but few people want to live in that funny little town

that together with *funny little town* forms a referential whose referend is *Podunk*.

the x is sometimes used to refer to an arbitrary member of a set, as e.g. in *the whale is a mammal*. In this 'generic' use, *the x* is of category N, hence *the* is of category (N;_N). Thus *the* in *the whale is a mammal* is of the same category as *a* in *a whale is a mammal*. Of course, the category assignment depends on the total context; *the* is assigned to ((N;N);_N) if *the whale* refers to a particular whale mentioned elsewhere in the text, since *the* forms a referential in that case. Thus we may distinguish *the* as a REFERENTIAL FORMATIVE from *the* as a GENERIC FORMATIVE by means of different category assignments.

5.1.7 Referentials such as *it*, *this* and *that* often have referends other than noun phrases:

```
(75) Lead floats in mercury. This  [seems strange].
             S              (N;S)    (S;N_)
             1               2 1      3 2
```

In its role as a functor in (75) *this* takes an argument of category S and forms a phrase of category N. In such cases *this* functions as a nominalizer of its referend. The referential *that* performs a similar function in *The potential difference was increased and that produced a reaction*. Likewise, *it* in *Boys who tease girls enjoy it* nominalizes its referend *tease girls*. Referentials may also involve a change in PERSON (see (76-d) e.g.). H. Hiż, in his paper *Referentials*,[5] discusses the fact that it is not always possible to replace a referential by its referend without ADJUSTING the grammatical category of the referend. Thus we would not say *The potential difference was increased produced a reaction*, but rather *Increasing the potential difference produced a reaction*. This adjustment of category involves transformations.

Certain occurrences of *do*, *does*, *did* and *so* function as referentials for predicate phrases:

(76) a. We *left* before you *did*
b. John *read a book* and *so did* Mary[6]
c. *Fry the chicken* like mother *does*
d. Mary *practices yoga* when I *do*

One possible analysis is illustrated by (77):

```
(77) Mary [practices yoga] when     I do
      N      (S;N_)     (S;S,S') N ((S';N);(S;N_))
      1      32 1        6 2 5   4    5  4  3
```

[5] In *Semiotica (1969), 136-166.*

[6] If the whole phrase *so did* is taken as a referential in (b) and *did* is assigned to the same category as *do* in (77) (let this category symbol be abbreviated by Y) then *so* is of category (Y; Y). On the other hand *so* may be regarded as the referential element and *did* as simply the tense carrier. Another possibility is to group *so* together with *and* forming a conjunction, and to regard *did* as the sole referential element (as well as tense carrier).

The referential *do* together with its referend *practices yoga* forms the referential phrase *practices yoga ... do*. The latter is a functor which, together with *I*, forms the referentially extended phrase *practices yoga ... I do*. This 'sentence-like phrase', which has been assigned to a category S′, is related to *I practice yoga* in that (77) can be paraphrased by:

(78)	Mary	[practices yoga]	when	I	[practice yoga]
	N	(S;N_)	(S;S_S)	N	(S;N_)
	1	2 1	6 2 5	4	5 4

which contains *I practice yoga* where (77) contains *I do*. In this sense the role of the sentence-like phrase *practices yoga ... I do* in (77) is the same as that of *I practice yoga* in (78).

The concept SENTENCE-LIKE PHRASE is not to be confused with that of a DEFORMED SENTENCE as used by Zellig Harris (see his *Mathematical Structures of Language* p. 73 for a discussion of sentence deformations). A 'deformed' sentence is the result of an operator φ_s operating on a sentence. E.g. *that* operating on *the sun rose* yields *that the sun rose*. But a sentence-like phrase is not formed in this way; rather it results from argument sharing or functor sharing, and it is not a nominalization of a sentence. The foregoing discussion suggests that the condition for membership in S′ might be given in the following form:

(79) Let s′ be an occurrence of a phrase which is not of category S or N in a text T′ and let T″ be the smallest text included in T′ and including s′. If s′ is sentence-like, there is a text T containing a phrase s of category S such that (i) T is a paraphrase of T″ in the context of T′, (ii) s′ does not occur in T, (iii) s does not occur in T″ and (iv) there is a part of s′, say x′, which is a shared element (in the context of T″) such that (a) x′ occurs in s without being a shared element there, or (b) x, which is like x′ except for number and person of a verb contained in it, occurs in s without being a shared element there.

To illustrate, taking the text (77) as T′ we have T″ = T′,

s′ = *practices yoga ... I do,* s = *I practice yoga,* x = *practice yoga,* and x′ = *practices yoga.* In this case the sentence-like phrase results from the use of the referential *do*, but not all sentence-like phrases involve referentials. For example, in *I learned yoga before Mary* and *John likes onions, Mary garlic* the sentence-like phrases are *learned yoga ... Mary* and *likes ... Mary garlic* respectively. Examples such as these will be treated in 5.2.

Sentence-like phrases function in the grammar much like those phrases which have thus far been assigned to S. Segmental functors which take arguments of category S also take sentence-like phrases as arguments (e.g., in (77) one argument of *when* is of category S′). Rather than establish a separate category S′ as indicated in (79), the policy here will be to assign sentence-like phrases to the category S. The assignment of intonation will determine which members of S in a given text form sentences (see 6.2).

5.2. ZEROING

5.2.1 One of the paraphrastic transformations in Zellig Harris's theory of transformations is called ZEROING. The operation of zeroing does not consist of an arbitrary deletion of words from a sentence, for as Harris states:[7]

> zeroing is carried out only if specified syntactic conditions are satisfied, and the zeroed words are recoverable (up to synonymity) from the remaining sentence.

As an illustration, the sentence:

(80) Richard plays piano and sings

would be considered a transform, via zeroing, of:

(81) Richard plays piano and Richard sings

where both occurrences of *Richard* in (81) denote the same person. According to rule (60), section 5.1.4, the second occurrence of *Richard* in (81) can be regarded as a referential whose referend is the first occurrence. Using subscripts to mark different occurrences, we may say that in (81) $Richard_1$ is the argument of *plays*

[7] Harris, *Mathematical Structures of Language*, 67.

piano and, indirectly, of *sings*, since $Richard_1$ is an argument of $Richard_2$. This is shown graphically in (81′):

(81′)

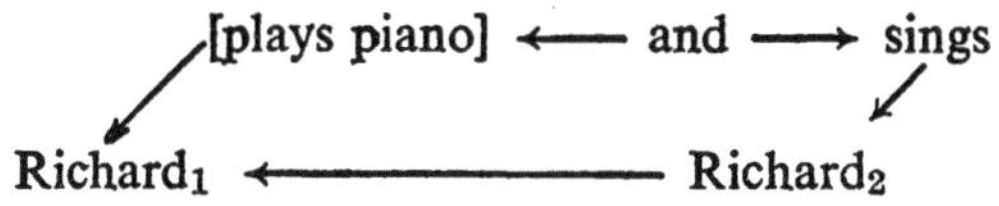

$Richard_2$ may also be replaced by unstressed *he*:

(82)

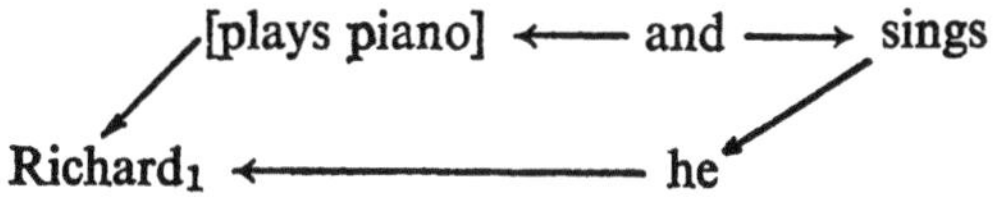

In (80) the single occurrence of *Richard* may be taken directly as argument of *sings*:

(80′)

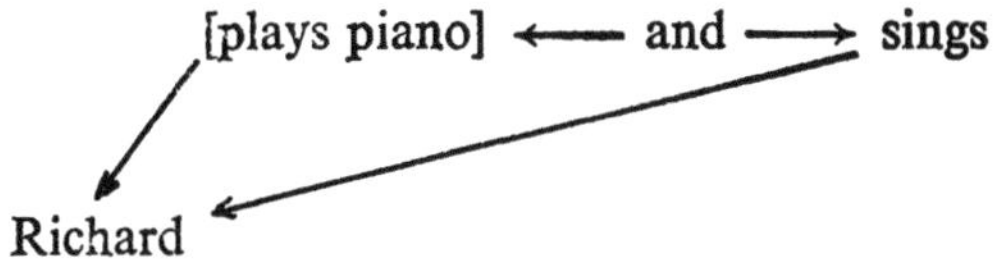

The structure assigned to (80) is therefore:

```
(83)      Richard [plays piano] and      sings
             N        (S;N_)    (S;S_S) (S;N)
             1         2 1       4 2 3   3 1
             1
```

Many cases of zeroing can be explained similarly as examples of argument sharing:

```
(84)
[Everyone sought] but      [nobody found] a peaceful solution
     (S;N)        (S;S_S)     (S;N_)               N
      2 1          4 2 3       3 1                 1
                                                   1
```

```
(85)    Either  you pass       or fail        the exam
        (S;S,S)  N  (S;N_,N) * (S;N,_N)          N
                 1    1 2        1 2             2
                 1                               2
```

(The category symbol for *either-or* is placed under *either*, hence the asterisk under *or*.) *pass* and *fail* share both of their arguments. The graph for (85) is:

(85′)

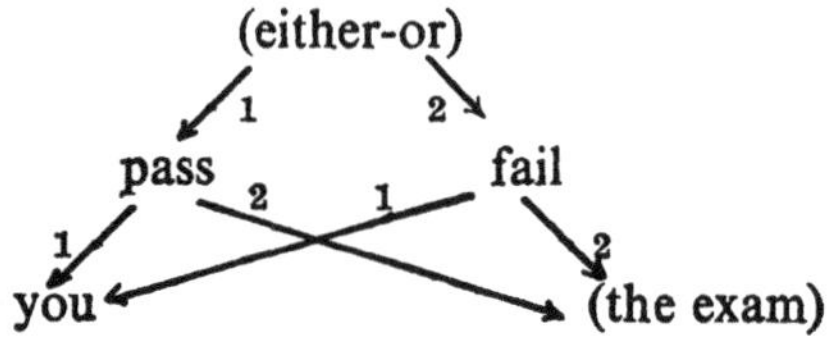

Sentences which are considered to be derived by zeroing of a VERB PHRASE may also be analyzed in terms of shared elements. In such cases the shared element is a functor and this leads to a new complication. But before taking up the subject of functor sharing, note that two kinds of argument sharing have been illustrated in this chapter. Examples (83) – (85) illustrate sharing in which an occurrence of a phrase x is an argument of two functors f and g. (In general, there may be n functors sharing the same argument.) Another type of argument sharing is illustrated by (77): it is not the entire first argument of *when* (i.e. *Mary practices yoga*) which is shared, only a part *(practices yoga)*. This will be called PARTIAL ARGUMENT SHARING. Partial argument sharing takes place when:

(86) a phrase x is an argument of a functor f, the phrase formed by this combination of x and f is an argument of a functor g, and f by itself is an argument of a functor h.

Thus in (77) *Mary* = x, *practices yoga* = f, *when* = g, and *do* = h. Note that if the argument of h is x rather than f, sharing is total since x is then the entire argument of both f and h. E.g. substitute *she is alone* for *I do* in (77) and let *she* = h.

5.2.2 *Functor sharing and zeroing*

Verbs may be 'zeroed' as well as nouns. Thus the second occurrence of *existed* is optional in (87):

(87) Dinosaurs existed before men (existed).

If the second occurrence of *existed* is omitted, its role is taken over by the first occurrence. Both *dinosaurs* and *men* are then arguments of the single occurrence of *existed*; but each is a FIRST argument:

(88)

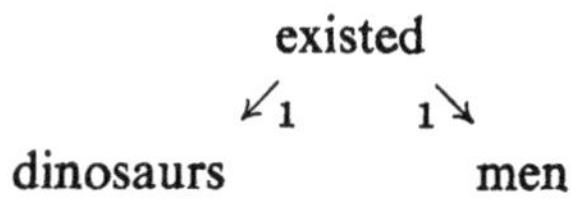

In (88) *existed* is not a two-argument functor, but a ONE ARGUMENT FUNCTOR USED TWICE. This is indicated by labeling both arcs with the numeral 1. If these arcs were labeled 1 and 2 instead, *existed* would be represented as a two-argument functor. The structure assigned to *Dinosaurs existed before men* on the basis of (88) is therefore:

(89)

```
Dinosaurs existed before  men
    N     (S;N_) (S;S,S)  N
    1       2 1    5 2 4  3
          (S;N)
           4 3
```

The second argument of *before* is the sentence-like phrase *existed ... men*. (If S′ is used for sentence-like phrases, the occurrences of S tagged by 4 are replaced by S′.) Following the usual procedure for constructing a graph from the structured text, the matchings indicated by the tags 2 and 4 each requires an arc from *before* to *existed*. The introduction of these PARALLEL ARCS means that an S-GRAPH (or multigraph) is being constructed. In an s-graph two vertices may be connected by as many as s parallel arcs. (89) requires a 2-graph:

(90)

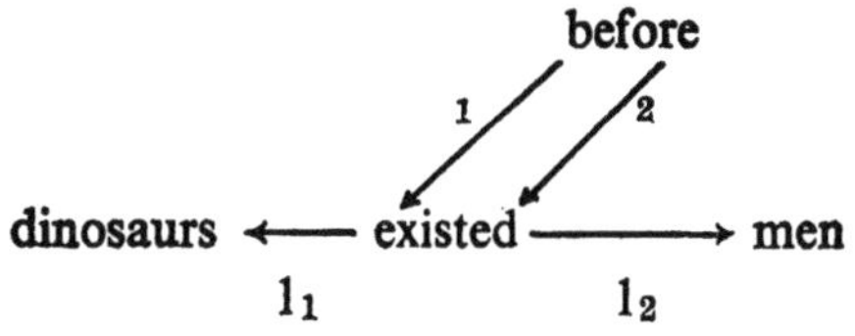

The labels 1_1 and 1_2 indicate that *dinosaurs* and *men* are in the first and second argument branches respectively of *before*. (Multigraphs have been used in the study of molecular structure, with vertices representing atoms and edges representing valence bonds – as illustrated in 3.2.)

From a transformational point of view the sentence:

(91) Pierre likes cold beer but wine at room temperature

results from zeroing an occurrence of *Pierre likes* after *but*. The phrase *Pierre likes* can be taken as a complex functor (since *x likes y* can be interpreted as *x likes* predicated about *y*) and we may say that this functor is shared. Then, without appealing to zero, the assigned structure is:

(92)

```
[Pierre likes] cold    beer but     wine [at room temperature]
   (S;N)     (N;_N)  N  (S;S,S)  N            (N;N_)
    3 2        2  1  1   7 3 6   4             5 4
    6 5
```

The sentence *John likes onions, Mary garlic* may be analyzed with the help of functor sharing as in (93):

```
(93)         John likes     onions Mary garlic
               N  (S;N_N)     N     N     N
               1   3 1 2      2     4     5
                  (S;NN)
                   6 4 5
```

(93′)

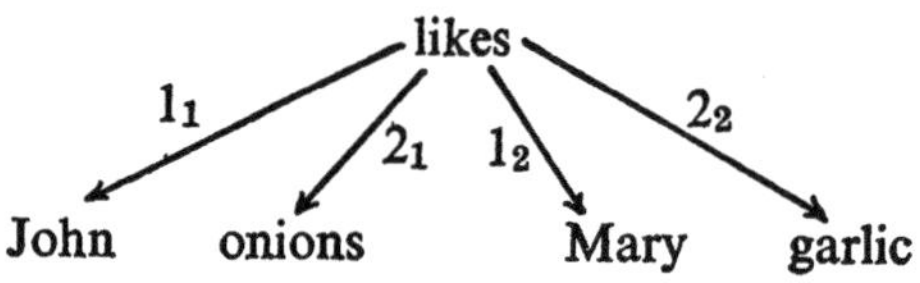

If the connective *and* were supplied, the arcs labeled 1_1 and 2_1 would be in the first argument branch of *and* while those labeled 1_2 and 2_2 would be in the second argument branch of *and*.

Certain occurrences of *and* and *or* form noun phrases out of noun phrases:

JOHN AND MARY *help linguists a great deal.*
The student body is composed of MATTER AND ENERGY.
Does Ethyl want A NEW FUR COAT OR A DIAMOND RING?

In such sentences *and* and *or* may be assigned to (N;N_N):

(94) Fred and Arthur [teach dancing]
N (N;N_N) N (S;N_)
1 3 1 2 2 4 3

With the help of functor sharing these occurrences of *and* and *or* can be treated as connectives between sentences (or sentence-like phrases). E.g. if *and* connects *Fred ... teach dancing* and *Arthur ... teach dancing* then *teach dancing* is a shared functor and instead of (94) we get (95):

(95) Fred and Arthur [teach dancing]
N (S;S,S) N (S;N)
1 5 3 4 2 3 1
4 2

If x and y are the noun phrases and z the predicate phrase in the preceding examples, the two analyses may be compared graphically as follows:

(96)

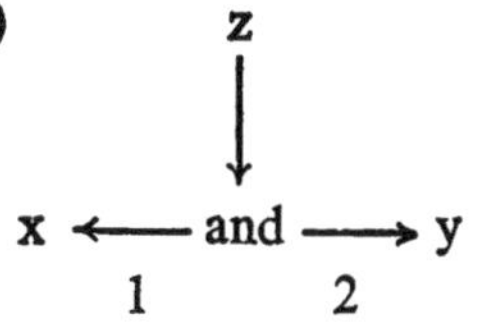

(97)

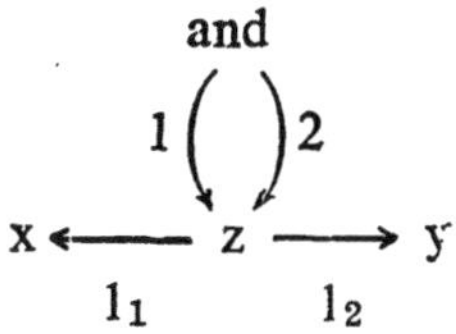

Both analyses will be accepted here. One structure shows the role that *and* plays in combining noun phrases while the other reveals its sentence connecting status. The relation between these two categories of *and* will be discussed further in chapter 7.

and and *or* may also combine adjectives to form adjective phrases. Consider *Mary bought a blue and white coat*:

(98)
[Mary bought] a blue and white coat
(S;_N) (N;_N) (N;N) ((N;_N);(N;N)_(N;N)) (N;N) N
6 5 5 4 2 4 3 2 1 1 3

This is not equivalent to *Mary bought a blue and a white coat* in which *and* combines noun phrases:

(99)

[Mary bought]	a	blue	and	a	white	coat
(S;_N)	(N;_N)	(N;N)	(N;N,N)	(N;_N)	(N;_N)	N
7 6	5 4	4 1	6 5 3	3 2	2 1	1
						1

Since (99) is paraphrased by *Mary bought a blue coat and Mary bought a white coat* the following structure may be assigned in place of (99):

(100)

[Mary bought]	a	blue	and	a	white	coat
(S;_N)	(N;_N)	(N;N)	(S;S,S)	(N;_N)	(N;_N)	N
4 3	3 2	2 1	8 4 7	6 5	5 1	1
(S;N)						1
7 6						

Functor sharing occurs in the examples given so far when a single verb is used in forming two sentences. In the case of a connective between verbs, since there are two verbs present functor sharing is not required to form conjoined sentences (or sentence-like phrases). This is illustrated in (101) and (102). (For convenience, V will be used to abbreviate some occurrences of (S;N_N).)

(101)

We	planned	and	executed	the attack
N	V	((S;N_N);V_V)	V	N
3	1	5 3 4 1 2	2	4

(102)

We	planned	and	executed	the attack
N	(S;N,N)	(S;S,S)	(S;N,N)	N
1	3 1 2	5 3 4	4 1 2	2
1				2

The corresponding graphs are:

(101′)

we ← [planned ← and → executed] → the attack

(102')

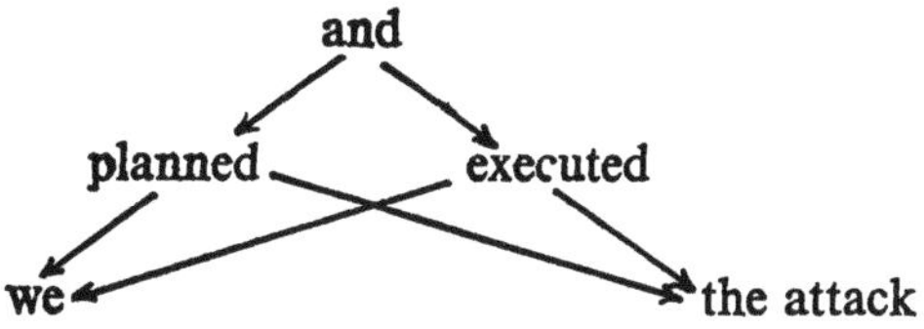

There are two occurrences of argument sharing here, but no functor sharing.

5.2.3 Not everything which is described as zeroing in a transformational grammar can be represented as argument sharing or functor sharing. E.g. if a sentence of the form *S*? is considered to be derived from *I ask you S* by zeroing the performative operator, this will not appear in the functor-argument structure of *S*?. The performative is simply not present in *S*?. The relation between the two sentences must be shown by means of transformations.

It appears that some, but not all, of the phenomena which are usually described in terms of transformations can also be described in terms of the functor-argument structure of a text without appealing to transformations. On the other hand, as a result of taking a multistructure point of view (see 4.2) the role of transformations may actually be expanded. These statements are somewhat ambiguous since transformations are defined in a different manner here than elsewhere. That definition will be given in chapter 7.

6

SUPRASEGMENTAL FUNCTORS

6.1. EMPHATIC AND CONTRASTIVE STRESS

6.1.1 *Emphatic Stress and Implied Contrast*

Emphatic and contrastive stress are frequently used in speech and easily identified by native speakers.[1] Emphatic stress may be applied to almost any word in a text:

(1) a. JACK *plans to climb the mountain*
b. *Jack* PLANS *to climb the mountain*
c. *Jack plans* TO *climb the mountain*
d. *Jack plans to* CLIMB *the mountain*
e. *Jack plans to climb* THE *mountain*
f. *Jack plans to climb the* MOUNTAIN

[1] In *An Outline of English Structure* (p. 52) Trager and Smith list five procedures for emphasizing any part of an utterance by means of stress, including change of pitch as well as loudness. Regarding 'normal' placement of stress, they state (p. 75): "the primary stress of a phonemic phrase will come as near the end as possible; here 'as possible' means that some items such as pronoun objects, certain adverbs, prepositions and others, do not have primary stress though they are normally the last thing in a phrase, and they get primary stress only with the shift morpheme". We may also include among unstressables occurrences of transformational *it* as in *It was Monday that I was sick.* In *Intonation, Perception, and Language* (p. 146) Lieberman writes: "We shall use the term 'emphasis' to identify instances where the distinctive feature [+ P_s] produces extra prominence on the vowel or vowels of a word (and its consonants), apart from the stress that the vowels of the word would have received from the phonologic stress rules. In other words, emphasis is prominence that is not predicted from the stress cycle. It may result from the presence of an emphatic morpheme in the underlying deep phrase marker."

Stressing *to* in the infinitive, as in (c), is probably contrastive. (Henry Hoenigswald has suggested the following example as a case in point: *I like reading but there is nothing* TO *read.*) Stressing *the*, as in (e), is tantamount to replacing *the mountain* with a proper name (of some mountain), presumably distinguished in the discourse.

Emphasizing a word by means of stress usually suggests a contrast – perhaps this is always true. But even though contrast is implied by stress on a phrase x, there may not be, overtly, any phrase y in the same text which serves as the other element in the contrast. The policy adopted here will be to use the term *contrastive stress* ONLY IF THE PHRASES BEING CONTRASTED ARE ACTUALLY PRESENT IN THE TEXT; otherwise extra stress on a single phrase will be referred to as *emphatic stress.* (In the case of contrastive stress it is not necessary that all of the contrasting phrases carry extra stress, but at least one must. This will be explained later. Of course, as pointed out above, all of the contrasting phrases must occur in the text.)

This distinction between contrastive and emphatic stress may seem arbitrary since contrast may be implied even when one member of a contrast is not present. My own feeling is that so-called emphatic stress IS contrastive; hence the above policy is intended to separate overt from implied contrast. The latter is more difficult to work with because of the problem involved in specifying the implicit phrase. I will deal with that problem in the following paragraphs. But first let us note two facts about emphatic stress.

The effect of emphatic stress is sometimes the same as inserting words like *indeed, certainly* or *really* into a sentence: (John IS a cad. John really is a cad. John is indeed a cad.) It is in such cases that the stress appears to be purely emphatic. The effect of emphatic stress on paraphrase relations is also important to note. In the following sentences (b), but not (c), is a paraphrase of (a):

(a) JANE wears slacks to church
(b) It is Jane who wears slacks to church
(c) It is slacks that Jane wears to church

When emphatic stress clearly implies contrast, a phrase might be added to the given text to supply a contrasting element. Let us call the resulting text an AUGMENTED TEXT. The augmented text contains contrastive stress. Of course, the phrase which is added may not be uniquely determined by the original text. E.g. any of the following phrases may be added to (1-d) to produce an augmented text:

— *not* FLY *over it*
— *not* MOVE *it*
— *rather than* IGNORE *it*
— *instead of just* LOOKING *at it*

The total discourse may eliminate some possibilities and knowledge of the world may render others implausible. Consider the examples in (2):

(2) a. JOHN gave Mary roses (BILL didn't; her OTHER boy-friends didn't; NO ONE ELSE gave her any; ...)
b. John GAVE Mary roses (he didn't SELL them to her; instead of LENDING them to her; ...)
c. John gave Mary ROSES (not ORCHIDS; not MONEY; NOTHING ELSE; he had tried to find ORCHIDS; ...)
d. What kind of flowers did John give Mary? John gave Mary ROSES (not ORCHIDS; no OTHER flowers; ...)
e. Frank is in pain. His NOSE is broken (his NECK isn't; his BONES aren't; ...)
f. Sue's RIGHT arm is broken (not her LEFT one; her LEFT one is O.K.; ...)
g. Ladies never imbibe BEFORE breakfast (Only AFTER; but sometimes DURING; ...)

In (a) – (c) there are many possibilities for choosing a contrasting word although perhaps the most plausible candidates are from the same grammatical category in the augmented text as the stressed word in the original text. In (d) the context suggests that the added phrase should contain the name of a flower, but not necessarily. In (e) anatomical considerations point to certain parts of the body, and in (f) to the LEFT arm. In (g) the word *before* seems to call for *after* or *during* to provide contrast.

(2) shows that in order to construct an augmented text which supplies an overt contrast it is necessary to take account of the linguistic context, the extralinguistic situation and the ideosyncracies of the stressed word.[2] The degree of indefiniteness about what phrase is to be added to the original text may be lessened by taking these factors into account, but is still not reduced to zero. The concept of an augmented text must be refined. It is not satisfactory just to add an arbitrary phrase which supplies contrast. We will place the following restriction on the added phrase: AN AUGMENTED TEXT MUST PARAPHRASE THE ORIGINAL TEXT. Clearly the linguistic context and extralinguistic situation have to be reckoned with at the very least.

The concept of an AUGMENTED TEXT WITH RESPECT TO EMPHATIC STRESS may be characterized as follows:

(3) Given a text T containing a phrase x with emphatic stress, an augmented text for T with respect to x is a text T′ such that (i) T is a part of T′, (ii) there is a phrase y in T′ such that contrastive stress exists between x and y in T′, (iii) the stressed word in y is not in T and (iv) T′ is a paraphrase of T in the context of the discourse in which T occurs.

The set of all such T′ for a given T will be symbolized by 'A(T)'. If A(T) is empty then we have a case of emphatic stress with no IMPLICIT CONTRAST.

An important factor in the relation between emphatic and contrastive stress is this oft mentioned but elusive contrast which is implicit in emphatic stress. In order to clarify this concept I propose the following definition (using the terminology of this section):

[2] From another point of view Henry Hoenigswald has suggested that sentences such as *Í am sick* have two kernels in their derivation in a transformational grammar, and Henry Hiż has pointed out that you cannot say *Í am sick* in every environment where *I am sick* is acceptable; i.e. the freedom of occurrence of the stressed form is less.

(4) Given a text T containing a phrase x with emphatic stress, THE CONTRAST IMPLICIT IN T DUE TO X is:

$$\{(u,v)| \quad u = T \quad \& \quad v \in A(T)\}.$$

If A(T) is empty then clearly there is no implicit contrast. Whether that situation will, in fact, occur is another matter.

6.1.2 *Contrastive Stress Functor*

Contrastive stress and emphatic stress are suprasegmental morphemes, features by which two texts may differ. The addition of extra stress on a phrase introduces an element of meaning. As pointed out earlier, emphatic stress affects paraphrase relations. Contrastive stress is a relational element which links two or more phrases in a text. Both of these suprasegmental morphemes will be taken as functors in this book.

A contrastive stress functor, symbolized by 'γ', may or may not be discontinuous, as seen in (5):

(5) a. Some people SEEM happy, some ARE
    b. Some people seem happy, some ARE

It is clear that *are* is contrasted with *seem* in (b) even though extra stress occurs only on *are*. (5-b) is an example of CONTRASTIVE stress since both of the phrases which are contrasted are actually present in the text. THE FACT THAT (b) IS A PARAPHRASE OF (a) AND IS IDENTICAL EXCEPT FOR EXTRA STRESS ON *seem* IS SUFFICIENT TO IDENTIFY THE FIRST ARGUMENT OF γ IN (b).

It is difficult to read (6) without stressing *are*:

(6) Some people SEEM happy, some are

But if *happy* is repeated,

(7) Some people SEEM happy, some are happy

then, although the acceptability may be lower than in (5), (7) is not as awkward as (6). It has been suggested that when two phrases are contrasted (by γ), extra stress is optional in the phrase which occurs earliest in the text, obligatory in the other. (8) shows that this is not always true:

(8) YEHUDI plays violin; Fred plays piano

Even though *Fred* is unstressed, it is clear that the contrast is between *Yehudi* and *Fred*; and (8) is an acceptable text.

Contrastive stress relates two or more phrases in a text. (9) illustrates three-way contrast:

(9) a. We WANT peace; we are PROMISED peace; why can't we HAVE peace?
  b. ... government OF the people, BY the people and FOR the people

Frequently two occurrences of contrastive stress in the same text are 'interlaced':

(10) FRED likes CUTE girls; BILL likes SMART girls

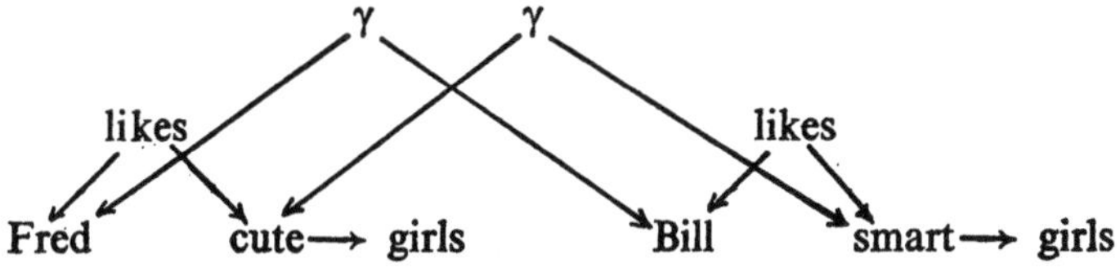

The arguments of γ will be ordered according to their order of occurrence in the text (strictly speaking, the order of occurrence of the stressed words). Thus *Fred* is first argument and *Bill* second argument of one occurrence of γ in (10) while *cute girls* is first argument and *smart girls* second argument of the other occurrence of γ.

The arguments of γ are not just the stressed words, but the phrases which these words dominate (see 4.6 for definition of *dominate*). In other words, I claim that in (10) the contrast is between *cute girls* and *smart girls*. Semantically, we are contrasting things that Fred and Bill like, namely, cute girls and smart girls.

If the arguments of γ are the phrases dominated by the stressed words, rather than the stressed words alone, then some cases where contrast has been described as existing between elements of

DIFFERENT grammatical categories may now be described as contrast between phrases of the SAME grammatical category:

(11) JOHN gave her roses. The trinkets were from OTHER people.

The stressed words in (11) are of categories N and (N;_N), but the contrasted phrases *John* and *other people* are of the same grammatical category. It should also be noted that just because two words in the same sentence carry extra stress, this does not mean that those words are part of the same contrast. That is evident in (10) and in the following example:

(12) The BASIC sin IS selfishness. (In reply, e.g. to the question "Is selfishness really such a big sin?")

In (12) *basic* and *is* are not both part of a single contrast relating phrases of different categories. What we have here is two separate occurrences of emphatic stress.

Phrases such as FRED ... BILL and CUTE *girls* ... SMART *girls* in (10) will be referred to as CONTRASTIVE STRESS PHRASES. A contrastive stress phrase consists of an occurrence of γ together with its arguments. The symbol 'C' will be used to designate the category of contrastive stress phrases; hence γ is a functor which, together with its (segmental) arguments forms a phrase of category C. For a particular occurrence of γ each stressed word is either an argument of γ or dominates an argument of γ. Thus in (10) *Fred* and *Bill* are arguments of one occurrence of γ while *cute* and *smart* dominate arguments of the other occurrence of γ. This is shown in (13):

(13)
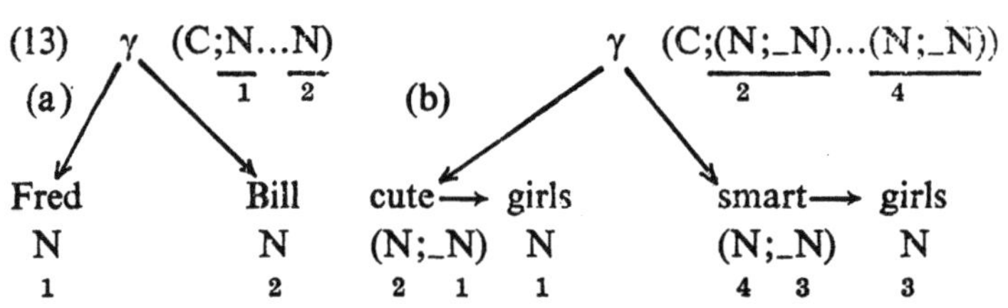

I have placed the position indicator '____' under the N's in (a) since stress occurs simultaneously with *Fred* and *Bill*. Likewise,

in (b) the position indicator is placed under '(N;_N)' since stress occurs simultaneously with *cute* and *smart* which belong to the category (N;_N). But there is a complication in (b): the arguments are *cute girls* and *smart girls* which are of category N. This is indicated by the numerals 2 and 4 which are placed directly under the N's on the left of the semicolons in the symbols:

$$\underset{2}{\underline{(N;\_N)}} \quad \text{and} \quad \underset{4}{\underline{(N;\_N)}}$$

This symbolism has the advantage of showing the category of the word which is actually stressed in the phrase. It is not necessary, however, since we know that the stressed word is the one which dominates the argument phrase. Therefore we may replace the category symbol in (13-b) by:

$$(C;\underset{2}{N} \ldots \underset{4}{N})$$

with the understanding that the 'position' of γ coincides with the words which dominate the noun phrases.

The structure assigned to the text *Some people* SEEM *happy*; *few* ARE *happy* is given in (14) with γ placed above *seem*, which dominates the first argument of γ:

(14)

$$\overset{\gamma}{(C;\underset{4}{\underline{(S;N\_A)}} \ldots \underset{7}{\underline{(S;N\_A)}})}$$

Some	people	seem	happy	few	are	happy
(N;_N)	N	(S;N_A)	A	(N;N)	(S;N_A)	A
2 1	1	4 2 3	3	5 1	7 5 6	6
	1					

(14′)

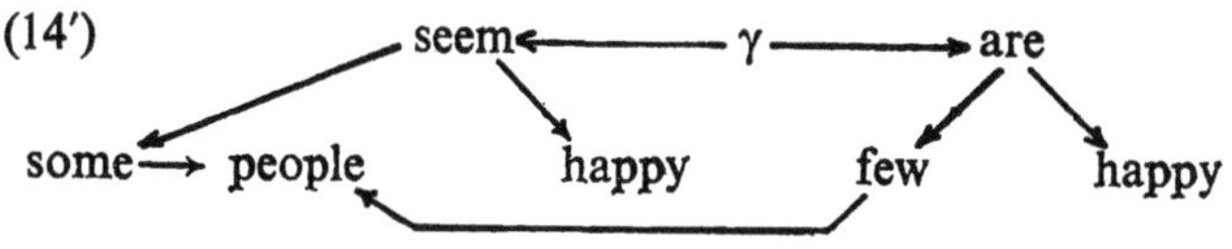

The category symbol for γ in (14) may also be written:

$$(C;S,S)$$

6.1.3 *Emphatic Stress*

The EMPHATIC STRESS FUNCTOR will be symbolized by 'e'. The phrase which this functor forms when applied to a segmental argument of some category X will be called an EMPHATIC STRESS PHRASE (esp), and the category of this esp (a subcategory of X) will be symbolized by $\acute{X}$. Hence:

(15) $$\text{g.c. (e)} = (\acute{X};X) \qquad (\acute{X} \subset X)$$

In the sentence LARRY *broke the window*, the esp is LARRY and it belongs to the category $\acute{N}$:

(16)

LARRY	broke	the window
$\acute{N}$	(S;N_N)	N
1	3 1 2	2

Instead of capitalizing the stressed word we will write the symbol for the emphatic stress functor above that word:

(17)

e($\acute{N}$;N)		
2 1		
Larry	broke	the window
N	(S;N_N)	N
1	4 2 3	3

In 6.1.2 the arguments of γ were taken to be the entire phrases dominated by the stressed words, not just the stressed words alone. I will argue that emphatic stress behaves the same way – that WHEN A WORD IS GIVEN EXTRA STRESS, THIS EMPHASIZES THE ENTIRE PHRASE WHICH THE STRESSED WORD DOMINATES.

If a speaker says:

(18) I prefer RED sports cars

he is emphasizing the color of sports cars, not the abstract notion of redness. An augmented text for (18) most likely REPEATS the phrase *sports cars* or contains a REFERENTIAL for it:

(19) a. I prefer RED sports cars to those which are NOT red.
b. I prefer RED sports cars, not BLACK ones or GREEN ones or BLUE ones or ...
c. I prefer RED sports cars to sports cars of OTHER colors.

If *sports cars* is not repeated in the augmented text and there is not a referential for it there, then it may be a SHARED ARGUMENT as in (20):

(20) a. I prefer RED sports cars to NON-red.
b. I prefer RED sports cars, not BLACK or GREEN or BLUE or ...
c. I prefer RED sports cars to OTHER colors.

In this case, however, the augmented text may be more or less awkward. In (20-a) and (20-b) sharing is rendered plausible by the acceptability of *non-red sports cars, black sports cars, green sports cars*, etc. But *other colors sports cars* causes some hesitation and, consequently, (20-c) may not be felt as a paraphrase of (18). It seems that when a phrase is added to (18) to form an augmented text, the added phrase either contains (i) *sports cars* or (ii) a referential for it or (iii) a functor which takes it as a shared argument.

Assuming that in (18) the entire phrase *red sports cars* is emphasized by the stress on *red*, the structure assigned to (18) is then:

```
(21)                      e
                        (Ń;N)
                         4 3

              I  prefer     red     sports cars
              N (S;N_N)   (N;N)         N
              1   1 4      3 2          2
```

(21′) I ← prefer → e → red → sports cars

The category symbol for e in (21) could also have been written $(\acute{N}_4;\underline{(N;\_N)}_3)$ to indicate that the stressed word *red* in the esp is of category (N;_N). It is simpler, however, to use the s

$(\underset{4}{\acute{N}};\underset{3}{N})$ with the understanding that the position of stress is on the word which dominates the noun phrase.

In example (17) the stressed word is an argument of a functor, but not a functor; in (21) the stressed word is a functor, but not the main functor; now we will consider stressing the main segmental functor in a sentence. If the claim that STRESSING A FUNCTOR f EMPHASIZES THE ENTIRE PHRASE DOMINATED BY f is true, then stressing the main segmental functor of a sentence should yield an EMPHATIC SENTENCE. In such a case e takes an argument of category S and forms an esp of category Ś. (Intonation may further affect the result; this will be discussed in the next section.) Consider the sentence *You* ARE *beautiful*:

(22) e$(\underset{4}{\acute{S}};\underset{3}{S})$

You are beautiful
$\underset{1}{N}$ $(\underset{3}{S};\underset{1}{N}\_\underset{2}{A})$ $\underset{2}{A}$

(22′) e
↓
are
↙ ↘
You beautiful

The role of e in (22) is like that of certain emphatic words such as *indeed*. Of course, when e creates an emphatic sentence e is anchored to the main segmental functor of that sentence, whereas *indeed* floats around:

Indeed you are beautiful
You are indeed beautiful
You are beautiful indeed

It is precisely in those cases where e is anchored to the main segmental functor of a sentence that the notion of contrast is least apparent. On the other hand, when a word other than the main segmental functor is stressed, contrast is usually suggested very strongly: *John likes* COLD *beer* (not WARM beer). This may be due to the relative ease with which a contrasting phrase can be tacked on to the sentence when the stressed word is not the main segmental functor.

But what is being contrasted with what in *You* ARE *beautiful*? Is it *You are* NOT *beautiful* or *You* WERE *beautiful* or *You* SEEM *beautiful* – or something else? This may be determined by the context. If not, at least the context should reduce the size of A(T) (see 6.1.1).

When the main segmental functor of a sentence is stressed the result is an emphatic sentence. (This does not preclude the presence of sentence intonation occurring independently along with e; for example, question intonation. That will be discussed in the next section.) The main segmental functor is not necessarily a verb:

(23) a. Unemployment increases AND inflation continues.
b. Call me IF you decide to go.
c. The customer can pay cash OR use the deferred payment plan.
d. Some people are bored BECAUSE they watch TV.

It is thus possible to distinguish emphatic conjunctions, emphatic conditionals, etc. And these are further affected by intonation patterns.

6.2. INTONATION

6.2.1 Intonation plays a fundamental role in speech: It may change an assertion to a question, register surprise, or convey subtle shades of meaning that defy classification. It is not learned at an advanced stage in life like the relative clause or pluperfect, but is part of a child's earliest utterances. An intonationless sentence is an abstraction unattainable in actual speech. Even a 'monotone' reading of a text is not intonationless. Hockett says of monotone speech:[3] "Such speech is not free from intonation. All PLs become /^{2}/ and /↑/ is replaced by /|/ or /↓ /, but other distinctions remain."

If one assumes that texts are STRINGS over some alphabet (see section 3.1 for a discussion of this point of view), then it is easy to see how intonation might be neglected in the study of syntax.

[3] Hockett, *A Course in Modern Linguistics*, 46.

Intonation does not fit neatly into the string concept since it forms another 'layer'; it may, of course, be represented by diacritics inserted among the elements of the strings. The point of view accepted in this book is that intonation may be taken as a morpheme[4] and that intonation morphemes enter into relations that fall within the domain of syntax.

It will be assumed here that in speech some intonation is always present. The concept of an intonationless phrase will be used, but the terms TEXT and SENTENCE, as applied to natural languages, will hereafter refer only to phrases which include intonation as well as segmental morphemes. The fact that an intonationless phrase is not to be counted as a sentence or a text does not imply that such a phrase is meaningless. Lack of intonation simply results in ambiguity. For example, the phrase *today is Monday* together with an appropriate intonation morpheme may convey a statement of fact, a question or an expression of surprise.

Intonation will be regarded as a functor whose argument is a phrase of category S. The members of S are intonationless phrases which may be thought of as 'sentence candidates'; i.e. if a phrase of category S serves as argument of an intonation functor, the resulting combination is a sentence. Obviously, not every phrase of category S in a given text forms a sentence; this matter will be taken up later. The interpretation of S as a category of intonationless phrases is consistent with the use of the symbol S up to this point since intonation functors have not yet been used in the analysis. (Note that $\acute{S} \subset S$; see 6.1.3.) The only difference is that now the term *sentence* will be applied only to those phrases which include intonation.

No special symbol will be introduced for the category of sentences. Instead, particular types of sentences will figure in the analysis: D, the category of declarative sentences; Q, the category of interrogative sentences (with subcategories Q_{yn}, yes-no question, and Q_{wh}, wh-question); I, the category of imperatives; and E, exclamatory sentences.

[4] See, e.g. Harris, *Structural Linguistics*, 45-58.

Different intonation functors are not necessarily represented by different intonation patterns. It is well known that the same intonation which indicates a declarative sentence in *He wants to go* is used in the interrogative sentence *Who wants to go?* The fact that different intonation functors may have the same 'spelling' should cause no greater consternation than the fact that different segmental functors frequently have the same spelling (e.g. *run* in *We run a grocery* and *Run to the grocery*).

An intonation functor which forms a declarative sentence out of a phrase of category S will be designated by a period placed after that phrase, hence the categorv of this 'declarative' functor is (D;$\underline{S}$).

Intonation is not usually assigned to conjoined or embedded phrases of category S separately. Thus there is only one intonation functor assigned in (24):

(24) When (Boris finished singing)
(S;_SS) S
3 12 1
(people who had been dozing applauded wildly) .
S (D;$\underline{S}$)
2 4 $\overline{3}$

However, there are some exceptions such as (25) and (26):

(25) You must have heard it. Or were you sleeping?
(26) I worked hard for what I have. And everyone should work hard for what he gets.

In this case there are two separate sentences and the second sentence begins with a conjunction. As an application of pure structural ambiguity such occurrences of a conjunction may be assigned to the category ((S;S_);_S). The segmental structure may then be represented schematically as in (27):

(27) x c y
S ((S;S_);_S) S
1 3 1 2 2

Applying sentence intonation to x and to cy, the following global structure results (taking (25) as example):

(28)

```
                 .                        ?
               (D;S)                  (Q;(S;S_))
                4 1                     6 5

            x        c           y
            S        ((S;S_);_S) S
            1         53 1     2 2
```

As we have already noted, an intonationless phrase is ambiguous. The assignment of various intonation functors to a phrase may result in different meanings. Thus the intonationless phrase:

(29) the wealthy heiress was opposed to a guaranteed annual income for all

may form either a declarative or an interrogative sentence, depending on the intonation pattern which is used. Of course, change in word order can also be used to signal the interrogative, but intonation must still be taken into account. Thus the same intonation pattern which forms an interrogative sentence out of (29) may also form an interrogative when the word order is changed so that *was* is placed in front:

(30) Was the wealthy heiress opposed to a guaranteed annual income for all ?
(Q;S)

But word order alone does not guarantee an interrogative sentence in (30). Even with the auxiliary fronted it is possible to produce an exclamatory sentence instead of an interrogative. Let '!' symbolize an exclamatory intonation functor. If a sentence is short, exclamatory intonation is easier to achieve in speech:

(31) Was she fat!
Can that girl dance!
Has this car got power!
Are those hard-hats messing up long-hairs!
Am I having trouble with this stretched out intonation pattern!

With some effort this same intonation pattern could be applied to the intonationless component of (30). Hence '?' is not superfluous in (30) in spite of the fronted auxiliary.

Emphatic stress is to be regarded separately from intonation.[5] E.g. in (32) emphatic stress separates (a) from (b), and (c) from (d), while intonation separates (a) from (c), and (b) from (d).

(32):

(a) Fred's looking for a job. (b) FRED's looking for a job.

(c) Fred's looking for a job? (d) FRED's looking for a job?

Example (33) illustrates the combination of emphatic stress functor and question intonation in the sentences ELAINE *is a dancer?* and *Elaine* IS *a dancer?*:

(33) (a)

$$\begin{array}{llllll} & e & & & & \\ & (\acute{N};N) & & & & \\ & {}_{2\ 1} & & & & \\ \text{Elaine} & \text{is} & & \text{a dancer} & & ? \\ N & (S;N\_N) & & N & & (Q;\underline{S}) \\ {}_{1} & {}_{4\ 2\ 3} & & {}_{3} & & {}_{5\ 4} \end{array}$$

(b)

$$\begin{array}{llllll} & & e & & & \\ & & (\acute{S};S) & & & \\ & & {}_{4\ 3} & & & \\ \text{Elaine} & \text{is} & & \text{a dancer} & & ? \\ N & (S;N\_N) & & N & & (Q;\underline{S}) \\ {}_{1} & {}_{3\ 1\ 2} & & {}_{2} & & {}_{5\ 4} \end{array}$$

(Note that in (a) *is* is assigned to the category (S;N_N) while its first argument is of category $\acute{N}$; this is permissible since $\acute{N} \subset N$. Likewise in (b) ? is assigned to $(Q;\underline{S})$ and its argument to $\acute{S}$ since $\acute{S} \subset S$. See 5.1.1, (15).)

Emphatic stress can also be placed in various positions in an EXCLAMATORY sentence: CAN *she dance*! *Can* SHE *dance*! *Can she* DANCE! This is another example of the need for separate functors representing emphatic stress and intonation.

[5] See, e.g. Harris, *Structural Linguistics*, 50-51.

The use of question words such as *where, when, why, what* does not render interrogative intonation superfluous any more than fronting the auxiliary did. In either case a change in intonation can change the type of sentence. The intonation pattern which indicates a declarative sentence in *John's a fink* yields an interrogative sentence in *What's a fink?* But a different intonation assigned to *what's a fink* may simply register surprise that someone has asked what a fink is. And still another intonation may indicate a reflective repetition by speaker B of a question by speaker A as B tries to think of an answer. With all the nuances available in the use of intonation, the categorization of sentences by means of intonation is no easy task. It is an open question as to what intonation functors are needed in a grammar in addition to those mentioned in this section.

In 5.1.3.2 we distinguished between restrictive and nonrestrictive relative clauses in terms of their segmental structures. In speech the difference is signalled by means of intonation. The pauses surrounding the nonrestrictive clause clearly separate the intonation pattern into two distinct parts, one interrupting the other. The intonation functor in this case may therefore be considered a two-argument functor, each argument consisting of a phrase of category S. This is shown in the following example:

(34) Bachelors who [are very discriminating]
N $(N;N_{-})$ $(S;N_{-})$
1 2 1 3 2
1
[are seldom satisfied] .
$(S;N)$ $(D;\underline{S},\underline{S})$
4 1 5 3 4

6.2.2 When a sentence is taken in isolation the intonation functor is coextensive with the segmental part of the sentence. On the other hand, when a sentence occurs within a larger text the intonation functor may not be coextensive with the entire segmental part of the sentence. This results from the fact that a functor in one sentence may have an argument in another. For example, in:

(35) That's Flo in the mini skirt. Look at her!

the second intonation functor (!) is coextensive with *look at her*, whereas the phrase which serves as argument of ! is *Flo ... look at her* (see the discussion of referentially extended sentences in 5.1). Of course, if the speaker who uttered (35) had simply nudged his companion and exclaimed *Look at her*! then the intonation functor ! would be coextensive with its argument. At any rate, the category symbols for intonation functors will be written $(D;\underline{\alpha})$, $(Q;\underline{\alpha})$, $(E;\underline{\alpha})$, etc. with the position indicator under the symbol α even when the intonation functor is not coextensive with the entire phrase of category α. (This is the same policy that was adopted for emphatic and contrastive stress where the argument of the stress functor is not just the word on which stress falls, but the entire phrase dominated by that word. See 6.1.3.) The part of its argument which is actually 'covered' by each intonation functor in a text may be determined with the help of the following rule:

(36) Let T_k be a structured text with successive intonation functors $f_1, ..., f_n$ and let x_r be the argument of f_r $(1 \leqslant r \leqslant n)$, where x_1 is an initial phrase[6] in T_k. Then f_1 covers x_1, f_2 covers that part of x_2 not also a part of x_1, f_3 covers that part of x_3 not also a part of x_1 or x_2, etc. (For each j $(1 < j \leqslant n)$, f_j covers that part of x_j which is not a part of any x_i such that $(1 \leqslant i < j)$.)

Of course, it is that part of x_r covered by intonation which is normally referred to as a sentence. We might call these parts INTONATION SENTENCES.

6.3. TEXTLETS

A text may contain many sentences. If a functor-argument graph is constructed to represent a particular assignment of structure

[6] There may be more than one initial phrase of the same category in T_k. E.g. in *Mary left when John arrived* both *Mary left* and *Mary left when John arrived* are initial phrases of category S, but only the latter serves as argument of the intonation functor.

to the entire text, the components of the graph may not correspond to individual sentences (see appendix for definition of COMPONENT). Referentials and contrastive stress often operate across sentence boundaries, hence an arc of the graph may connect phrases in different sentences. We will say that there is a SYNTACTIC LINK between two sentences (or any two phrases) in a structured text if a functor in one has an argument (or part of an argument) in the other, or if some functor has an argument (or part of an argument) in each of them. The resulting combination constitutes a syntactically linked part of the structured text. (This differs from the notion of a 'syntactically connected' phrase as defined by Ajdukiewicz or Bar-Hillel.) Let us use the term TEXTLET[7] to refer to the maximal syntactically linked units in a structured text. More precisely: let x_1 be any sentence in a structured text T_i; x_2 any other sentence in T_i syntactically linked to x_1; x_3 any sentence in T_i (other than x_1 or x_2) syntactically linked to x_1 or x_2; ... and x_n any sentence in T_i (other than x_1 or x_2 ... or x_{n-1}) syntactically linked to x_1 or x_2 ... or x_{n-1}, where $n \geqslant 1$. If no other sentence of T_i is syntactically linked to any of these, that part of T_i consisting of $x_1, ..., x_n$ is a textlet. Note that the subscripts do not indicate the order of occurrence of these sentences in the text.

Now suppose a graph G_i is constructed for T_i in the following manner: the vertices of G_i are the sentences of T_i, and a pair of vertices (x,y) forms an arc of G_i if and only if there is a syntactic link between the sentences x and y. (Of course, (x,x) forms an arc since there is obviously a syntactic link between a sentence and itself.) The components of G_i correspond to the textlets in T_i. However, G_i is not a functor-argument graph. In fact, the components of the functor-argument graph of T_i do not necessarily correspond to the textlets in T_i. This lack of one-to-one correspondence results from the presence of complex functors. For example, there are two textlets in the analysis of:

7 Zellig Harris has used this term in a somewhat different sense to refer to a sequence of sentences just long enough to contain certain distributional l imitations. (See Harris, "Eliciting in linguistics" [1953].)

(37) John bought two tickets to the opera. He gave one of them to Patricia. Mary was furious.

The first two sentences form one textlet and the third sentence forms another. Corresponding to the second textlet there will be one component in the functor-argument graph if the assigned structure is:

(38)	Mary	was	furious	.		was ←——.
	N	(S;N_A)	A	(D;$\underline{S}$)	↙	↘
	1	3 1 2	2	4 3	Mary	furious

but two components if the assigned structure is:

(39)

Mary	was	furious	.	Mary ← [was → furious] ←.
N	((S;N_);_A)	A	(D;$\underline{S}$)	
1	3 1 2	2	4 3	

since in (39) *was ←-furious* is itself a component of the graph. In the analysis of *Pablo likes very hot chili*, if *very* is to be interpreted as a modifier of *hot* then it is assigned to the category ((N;_N);_(N;_N)) since its argument is *hot*. The result is a complex functor *very hot*, and in this case there is no alternative which avoids the complex functor.

If a structured text T_i contains no complex functors, there is a one-to-one correspondence between the textlets in T_i and the components of the functor-argument graph of T_i. Otherwise, a textlet in T_i corresponds to a set of components in the graph, each complex functor introducing an additional component. In any event, textlets, not sentences, are the maximum grammatical units in the analysis proposed here.

7

RELATIONS BETWEEN STRUCTURED TEXTS

7.1. The point of view taken in the preceding chapters is that the structure of a text can be represented by treating some of the phrases in that text as functors and others as arguments of those functors. In a structured text T_i each functor-argument phrase (see 4.6) contributes something to the meaning of T_i. Presumably, the functor-argument phrases in T_i make sense to native speakers and help to make sense out of the text as a whole. It is assumed here that native speakers can render judgements as to whether two texts have the same or different meanings, but no attempt will be made to say what those meanings are.

It is often the case that when one text is a paraphrase of another many of the same functor-argument phrases are found in both – perhaps with some modifications. In this chapter we shall investigate the relation between functor-argument structures in paraphrastically related texts. If f is an occurrence of a functor in a structured text T_i and A_f is the set of phrases that occur in T_i as arguments of f, then A_f will be called the ARGUMENT SET of f in T_i. To illustrate this definition consider the argument set of each occurrence of *gave* in (1):

(1) (a) John gave bonbons to Mary.
(b) To Mary John gave bonbons.
(c) John gave Mary bonbons.

The argument set of *gave* in (a) is {John, bonbons, to Mary}, that in (b) is identical (since A_f is an unordered set), and the argument set of *gave* in (c) differs only in containing *Mary* instead of *to Mary*. Meaning is preserved in passing from one of these sentences to the other and the functor-argument phrase dominated by *gave* is 'carried over' with only slight modifications – modifications that can easily be formalized. This suggests that in a formalizable paraphrastic relation between two structured texts T_i and T'_i the preservation of meaning depends on some kind of preservation of the functor-argument phrases in passing from T_i to T'_i (or vice versa). These remarks are summed up informally in (2):

$$(2)\quad \underbrace{\ldots (f, A_f) \ldots}_{T_i} \qquad\qquad \underbrace{\ldots (f', A_{f'}) \ldots}_{T'_i}$$

The relation of f to f′ is frequently more complicated than in the pairs of sentences in (1) where f and f′ were both *gave*. E.g., if f = *printed* in (3)a, then f′ = *was printed by* in (3)b:

(3) a. The Times printed the story.
b. The story was printed by The Times.

Of course, the argument sets of f and f′ in (3) are identical.

The informal and somewhat vague notion of a meaning preserving relation between structured texts suggested by (2) needs to be made more precise. To this end let us define a relation between structured texts T_i and T'_i such that (A) T'_i is a paraphrase of T_i and (B) T'_i can be obtained from T_i by formal operations on the functors and arguments of T_i.

(A) is a semantic criterion, (B) a structural criterion.

(A) is EMPIRICALLY DECIDED. T and T′ need not be considered equally acceptable by native speakers. There may be a feeling that one is 'better English' than the other or that one is awkward or unusual, but both must be IN the language.[1] It follows that although meaning is preserved, no claim is made about preserving degree of acceptability. (This may allow for greater flexibility in dealing with forms that are becoming obsolete, new forms entering

[1] An expression which seems awkward or semantically bizarre may still be considered 'in the language' if native speakers can operate with it as they do with more normal expressions; i.e. if they can paraphrase it, draw consequences from it, nominalize it, etc.

the language, forms used only in special circumstances or forms characteristic of some sublanguage.) The paraphrase relation may hold not only between declarative sentences, but between interrogatives, imperatives and other texts as well.

(B) is to be FORMALLY STATED. Out of all possible paraphrases of a text there are some which bear an obvious structural relation to that text and others which do not. In the following example (ii) is related to (i) by a permutation, but although (iii) is a paraphrase of (i) there is no obvious structural relation between them:

(4) (i) Orion will be visible from my window tonight.
(ii) Tonight Orion will be visible from my window.
(iii) By looking through my window tonight one can see the group of stars called 'Orion'.

What we want is a small finite set of rules that will describe structural relations between paraphrastically related texts – rules consisting of formal statements in terms of the functor-argument structures of the texts. A list of such rules will be given in 7.2.

A relation which satisfies BOTH the empirical criterion (A) and the formal criteria suggested in (B) will be called a TRANSFORMATION. This relation is defined on structured texts. (The symbols T, T′, ..., T^n without subscripts represent texts; with subscripts, structured texts.)

7.2. FORMAL CRITERIA

The list of rules in this section provide formal criteria for (B), 7.1. Each rule is preceded by examples to illustrate the structural relation which the rule describes.

Rule (I) involves PURE STRUCTURAL AMBIGUITY (see 4.2). Numerous examples of pure structural ambiguity have already been given (e.g. chapter 4: (1)-(3), (4)-(6), (8)-(11) and (18)-(19)). In each case the difference between T_i and T_j is the result of a functor being assigned to a different category. Consider *sent* in *John sent Mary roses*:

```
(5) Ti: John sent          Mary roses .
        N    (S;N_NN)  N    N  (D;S)
        1     4 1  23  2    3   5 4

    Tj: John [sent           Mary] roses .
        N    ((S;N_N);_N)  N    N  (D;S)
        1      4 1 3   2   2    3   5 4
```

In T_i *sent* is a 3-argument functor; in T_j *sent* is a one argument functor which, together with *Mary*, forms the 2-argument functor *sent Mary*. The structural relation involved in (5) is generalized in rule (I). (Note: If x is an occurrence of a phrase in T_i then x′ is the corresponding occurrence of that phrase in T_j.)

(I) Let f be an occurrence of a functor in T_i and let the arguments of f in T_i be $a_1, \ldots, a_n$ ($n \geqslant 2$). If f is of category $(\beta; \gamma_1, \ldots, \gamma_n)$ in T_i, assign f′ to the category:
$((\beta; \gamma_1, \ldots, \gamma_{r-1}, \gamma_{r+1}, \ldots, \gamma_n); \gamma_r)$
in T_j, where $1 \leqslant r \leqslant n$.

Aside from the regrouping which the new category assignment entails, T_i and T_j are alike. The phrases $a_1, \ldots, a_n$ are of the same category in both T_i and T_j. The graph of T_j will contain a 'complex vertex' not present in T_i. If rule (I) is applied to T_j in (5), taking f to be *sent Mary* (of category (S;N_N)), the result is T_k:

```
(6) Tk: John [[sent Mary]  roses]
        N    ((S;N_);_N)   N
```

Rule (II) applies to cases of argument sharing which are equivalent to a certain type of 'zeroing':

```
(7) Ti: Fashions come  and     fashions go      .
        N        (S;N_) (S;S_S) (N;N)  (S;N_) (D;S)
        1         2 1    5 2 4   3 1    4 3    6 5
        1

    T′i: Fashions come  and     go     .
         N        (S;N_) (S;S_S) (S;N) (D;S)
         1         2 1    4 2 3   3 1   5 4
         1
```

(Note that the second occurrence of *fashions* in T_i has been assigned to (N;N) in accordance with (60), 5.1.4.)

(II) In T_i one occurrence of a phrase x is (part of) an argument of a functor f_1 and another occurrence of x is (part of) an argument of a functor f_2; one of these occurrences of x is deleted and the other is then shared as (part of) an argument of both f_1 and f_2.

E.g. in (7) x = *fashions*, f_1 = *come*, f_2 = *go*; the single occurrence of *fashions* in T'_i is an argument of both *come* and *go*. The qualification *part of* is necessary in (II) since either occurrence of x may itself have an argument elsewhere in the text.

Rule (III) concerns functor sharing:

(8) T_i:

007	arrived	before	the police	arrived	.
N	(S;N_)	(S;S_S)	N	(S;N_)	(D;$\underline{S}$)
1	2 1	5 2 4	3	4 3	6 5

T'_i:

007	arrived	before	the police	.
N	(S;N_)	(S;S,S)	N	(D;$\underline{S}$)
1	2 1	5 2 4	3	6 5
	(S;N)			
	4 3			

By deleting one occurrence of *arrived* from T_i and shifting its argument to the remaining occurrence, T'_i is obtained.

(III) T_i contains two occurrences of a phrase x (say x_r, x_t) which are functors of the same category α. One occurrence of x (say x_t) is deleted, resulting in T'_i such that
(i) the arguments of x_t become arguments of x_r which is assigned an additional category β in T'_i, and
(ii) β differs from α by virtue of the position of x_r with respect to its newly acquired arguments.

In (8) α = (S;N_) and β = (S;N); the difference between α and β reflects the difference in positional relation between functor and argument. In graphing (8) T'_i requires a multigraph and T_i does not.

In the next example T'_i is obtained from T_i by argument deletion. However, unlike the cases covered by rule II, no argument sharing results from the deletion:

(9) T_i: John wrote something .
N (S;N_N) N (D;$\underline{S}$)
1 3 1 2 2 4 3

T'_i: John wrote .
N (S;N_) (D;$\underline{S}$)
1 2 1 3 2

(10) T_i: You be good ι
N (S;N_A) A (I;$\underline{S}$)
1 3 1 2 2 4 3

(ι = imperative intonation functor)

T'_i: Be good ι
(S;_A) A (I;$\underline{S}$)
2 1 1 3 2

(IV) A functor f in T_i is replaced by a functor f′ in T'_i such that
(i) f and f′ are represented by identical phrases, but in different categories, and
(ii) $A_{f'} \subset A_f$ (proper inclusion).

The next rule, V, covers cases where a functor f in T_i is replaced by a functor f′ which takes exactly the same arguments, although in different order of occurrence, to form T'_i. This rule concerns permutations which are not accompanied by other changes:

(11) T_i: I like this .
N (S;N_N) N (D;$\underline{S}$)
1 3 1 2 2 4 3

T'_i: This I like .
N N (S;NN_) (D;$\underline{S}$)
2 1 3 2 1 4 3

(12) T_i: (The fur flies) when (Spiro speaks) .
S (S;S_S) S (D;$\underline{S}$)
1 3 1 2 2 4 3

T'_i: When Spiro speaks the fur flies .
(S;SS) S S (D;$\underline{S}$)
3 21 2 1 4 3

(V) A functor f in T_i is replaced by f′ in T'_i, where
(i) f and f′ are identical phrases, differing in category only by virtue of position relative to arguments, and
(ii) $$A_{f'} = A_f$$

A functor f in T_i may be related to a functor f′ in T'_i by certain constants such as *wh-*, *-ly*, *t(be)*, *it*, *what*, etc.

(13) music ← [which is soothing]   soothing → music
behaved ← [in a strange manner]   behaved ← strangely
[Newton invented] → gravity
[What Newton invented was] → gravity
The news ← startled → them
They ← [were startled by] → the news

In the last example, since *were startled by* is assigned to category (S;N_N) and *were startled* is of category (S;N_), this occurrence of *by* must be assigned to ((S;N_N);(S;N_)_). In the first three examples $A_{f'} = A_f$ but in the last example *them* becomes *they* in the passive.

A general rule covering relations of the type illustrated in (13) must include a list of structural changes involving various constants and grammatical categories. The list may be fairly extensive and will probably be found in one form or another in any transformational grammar. Rule (VI) is a broad classification of such changes.

(VI) A functor f in T_i is replaced by f′ in T'_i according to one of the formulas in (i):
(i) (Expressions which represent f and f′ are italicized; expressions for their arguments are also given, but not italicized; *t* is appropriate tense)
N *wh(N)t(be)A*   *A* N
V *in a(n) A manner*   V *Aly*

$N_1 V$ N_2	*What* $N_1 V$ *t(be)* N_2
$N_1 V$ N_2	N_2 *t(be) V-en by* N_1
etc.	

(ii) $A_f' = A_f$ except possibly for arguments which are pronouns in different case forms in T_i and T'_i.

The next rule involves referentials. Some occurrences of referentials can be considered as replacements for phrases which are repeated in a text (or 'repeated' with some modification). In the following examples (14-17) repeated phrases and referentials are italicized:

(14) If *Vladimir* is to be great, *Vladimir* must suffer.
If *Vladimir* is to be great, *he* must suffer.

(15) *Jane* is a feminist. *Jane's* husband is a male chauvinist.
Jane is a feminist. *Her* husband is a male chauvinist.

(16) Jane is *bellicose* and John is *bellicose.*
Jane is *bellicose* and John is *too.*

(17) *Jane smokes cigars. Jane's smoking cigars* bugs John.
Jane smokes cigars. This bugs John.

(In (14) and (16) the repeated phrase in T_i is not modified. In (15) the possessive suffix is added to *Jane* and in (17) *Jane smokes cigars* is nominalized.) These examples illustrate the REFERENTIAL REPLACEMENT rule:

(VII) The phrases x and x′ occur in T_i, where
(i) x′ is a repetition of x in the same category, or
(ii) x′ is the result of x combined with the possessive affix, or
(iii) x′ is a nominalization of x.

Replace x′ by ref (x′,T_i), a referential for x′ in the context T_i, to obtain T'_i.

(16) may be paraphrased by *John and Jane are bellicose*. In this sentence if *John* and *Jane* share the functor *are*, then number agreement will be lacking. On the other hand, if the whole phrase *John and Jane* is taken as an argument of *are* there will be number

agreement, but then *and* must be assigned to the category (N;N_N). The relation between structures in which *and* occurs in these two categories is illustrated in (18):

(18)

T_i:	Jane	is	bellicose	and	John	is	bellicose	.
	N	(S;N_A)	A	(S;S_S)	N	(S;N_A)	A	(D;S̲)
	1	3 1 2	2	7 3 6	4	6 4 5	5	

T'_i:	Jane	and	John	are	bellicose	.
	N	(N;N_N)	N	(S;N_A)	A	(D;S̲)
	1	3 1 2	2	5 3 4	4	

The corresponding graphs are:

(18′)

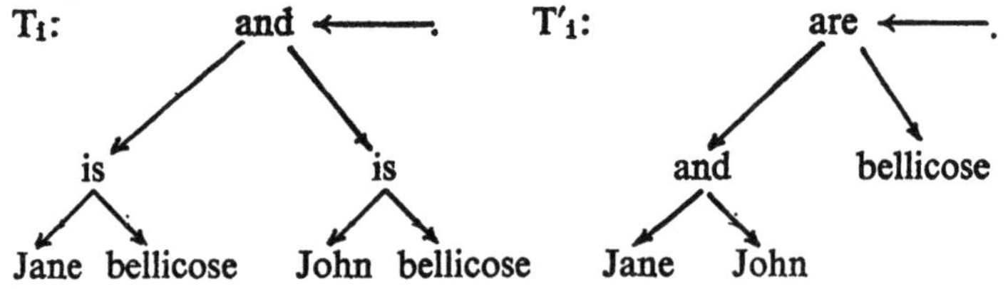

Zellig Harris has suggested the possibility of a 'distributive law' connecting such structures (in a seminar, U. of Pennsylvania, 1970). Using Harris's operator notation, the segmental parts of T_i and T'_i may be written:

T_i: and (is (Jane, bellicose), is (John, bellicose))
T'_i: are (and (Jane, John), bellicose)

The relation here is somewhat more complicated than the distributive law in arithmetic since *is* must be replaced by *are*. But in some cases no such change is required:

(a) Dogs are mammals and cats are mammals.
(a′) Dogs and cats are mammals.
(b) We saw the film and we saw the play.
(b′) We saw the film and the play.
(c) You believe and I believe.
(c′) You and I believe.

The graphs for the segmental parts of (c) and (c′) are:

(20)

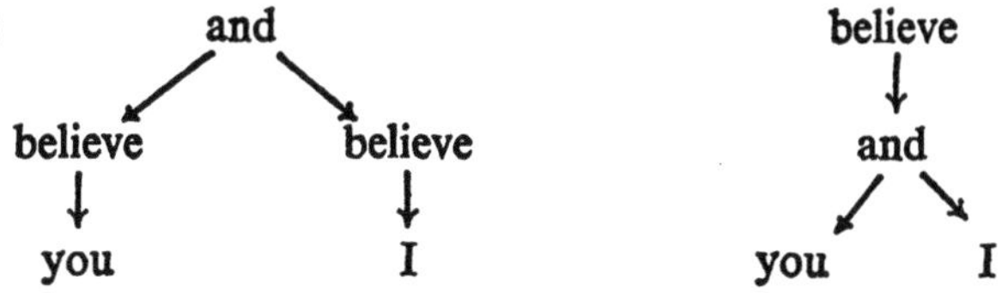

Note that if *is bellicose* and *are bellicose* are treated as complex functors of category (S;N_), then in place of (18′) we would have:

(21)

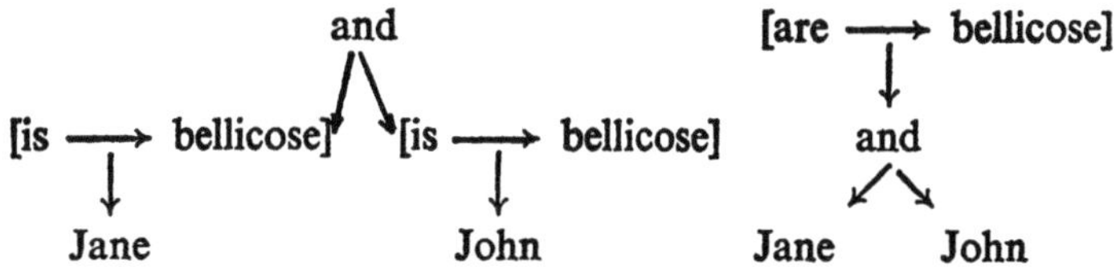

which is like (20) except for the complex vertices. Similar results may be obtained using *or* in place of *and*:

Is it black or is it white?
Is it black or white?

These examples suggest a relation which may be conveniently expressed by graphs (let x be the intonation functor):

(VIII-a)

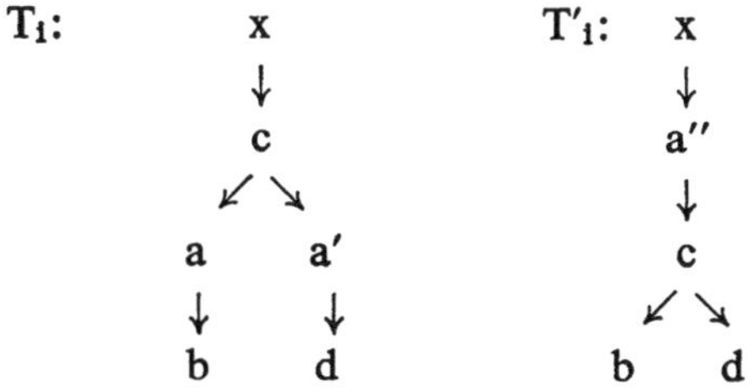

(i) c is either *and* or *or*;
(ii) c is of category (S;S_S) in T_i;
(iii) a, a′ and a″ may differ in number and person of verb, but not otherwise.

Verb phrases may also be conjoined, resulting in a complex functor of the same category as the conjoined phrases:

you ← [may → [win ← or → lose]]
we ← [buy ← and → sell] → used cars

(VIII-b) The segmental part of T_i is of the form:

x c y
S (S;S_S) S

where x and y contain verb phrases v_1 and v_2 respectively, v_1 and v_2 are of the same category (say V), and the argument(s) of v_1 in x are identical with the argument(s) of v_2 in y. The argument(s) of v_2 are deleted to form T'_i in which c is assigned to (V;V_V) and the argument(s) of v_1 become the argument(s) of the complex functor [v_1c v_2] (or [t[v_1c v_2]] where t = *can*, *will*, *may*, etc.).

The relation between contrastive and emphatic stress was discussed in chapter six with the help of the concept of an augmented text (6.1.1, (3)). This relation is illustrated by the following sentences:

SOME people were fooled.
SOME people were fooled, but not ALL.

Contrast is implied in the first sentence and explicit in the second (augmented) sentence. The corresponding structured texts are given in (22):

(22) T_i: e
(Ń;N)
3 2

Some people [were fooled] .
(N;_N) N (S;N_) (D;$\underline{S}$)
2 1 1 4 3 5 4

```
T'ᵢ: γ
     (C;N ... N)
      9 2    4

     Some    people [were fooled] but       not   all     .
     (N;_N)   N      (S;N_)   (S;S_S) (S;S) (N;N) (D;S̲)
      2  1    1       3 2      7 3 6   6 5   4 1   8 7
      2       1      (S;N)                   4
                      5 4
```

The graph for T'_i is:

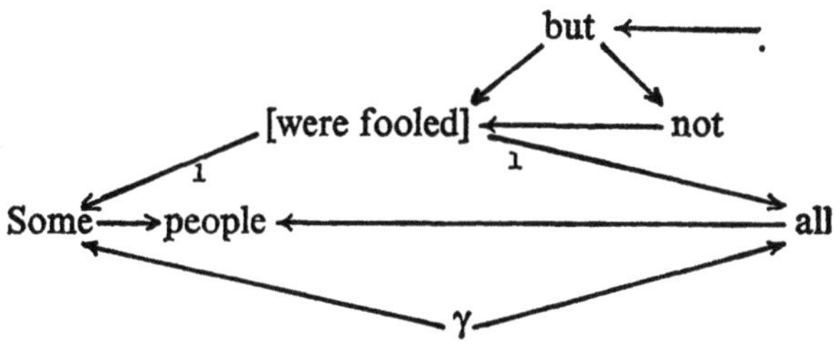

(*were fooled* is a one-argument functor used twice.)

In general, the structural relation illustrated by (22) can be stated as in (IX):

(IX) (i) e occurs in T_i and γ in T'_i;
(ii) if τ and τ' are the segmental parts of T and T' respectively, τ is identical with a part of τ', say τ^o;
(iii) the argument of e in T_i is identical with an argument of γ in T'_i, and their positions correspond in τ and τ^o; and
(iv) the words which dominate the remaining arguments of γ are not in τ^o.

Note: When the word which dominates an argument of γ is not in τ^o, other words in that same argument phrase may be in τ^o; e.g. in (22) *all* is not in τ^o, but *people* is. The word which dominates an argument of γ is usually, though not always, stressed (see 6.1.2, second paragraph).

Rules (I) – (IX) provide a tool for studying relations between texts in terms of the functor-argument structures of the texts. By extending the list of rules more relations may be taken into account, but the basic concept of a transformational relation (in this book) remains as described in 7.1. Rules (I) – (IX) are not, in themselves, transformations. The following definitions make this clear:

(23) T'_i is a DIRECT TRANSFORM of T_i if and only if
(A) T'_i is a paraphrase of T_i, and
(B) T'_i is related to T_i by one of the rules (I) – (IX).

(24) T'_i is a TRANSFORM of T_i if and only if there are $T^o{}_{j_o}$, $T^1{}_{j_1}$, ..., $T^n{}_{j_n}$ such that $T^o{}_{j_o} = T_i$, ..., $T^n{}_{j_n} = T'_i$ and $T^{k+1}{}_{j_{k+1}}$ is a direct transform of $T^k{}_{j_k}$ for each k ($0 \leqslant k < n$).

(If the list of rules (I) – (IX) is extended, then these definitions are extended accordingly.)

As an illustration of this definition, (26) is a direct transform of (25) by rule (V) and (27) is a direct transform of (26) by rule (VI), hence (27) is a transform of (25) according to (24):

(25) The guru titillated their fancy at the ladies club.
(26) At the ladies club the guru titillated their fancy.
(27) At the ladies club what the guru titillated was their fancy.

All the rules except (I) apply to two different texts. But in rule (I) just one text is involved. There is no 'zeroing', adding constants, replacing of a phrase by a different phrase, nor even a change in word order; only a change in grouping and a corresponding change in category assignments. (I) is unique in this respect. We may ask why such a change should be listed as a transformation.

The motivation for making different functor assignments corresponding to different interpretations within one 'reading' of a text was discussed in 4.2. An alternative would be to assign just one structure to a reading and then provide focusing rules which select one or another phrase in the text as 'topic'. Such focusing rules would then perform the same function as rule (I). However, as pointed out in 4.2, this would involve selecting one of many possible functor assignments as most appropriate. It is difficult to imagine a functor assignment which is 'neutral' with respect to the different interpretations: a particular choice would have to be justified on some grounds. The approach which has been taken is more in keeping with the general policy in this book of using functor assignments, as far as possible, to characterize structure. The terminological problem arising from different uses of the term *structure* in linguistics may be minimized here by referring to the set of functor assignments corresponding to a given reading of a text as the (composite) structure corresponding to that reading.

Transformations by rule (I), in spite of their unique character, are perfectly in accord with the general concept of transformation described in 7.1: If two interpretations fall within a single reading of a text then the paraphrase criterion holds; the formal criterion is clearly satisfied by (I). Regrouping of phrases is often required in linguistics. The definition of transformation given here covers certain regroupings which other definitions of transformation do not.

APPENDIX: TERMINOLOGY FROM GRAPH THEORY[1]

Roughly speaking, a graph is formed when each element of a set X is associated with various elements of the same set. The elements of X are the vertices of the graph, and each ordered pair consisting of a vertex and one of its 'associates' is an arc of the graph. An arc is represented by an arrow in the diagrams.

DEFINITION: Let X be a set and F a function such that $Fx \subset X$ for each $x \in X$. A graph (X,F) is the set:

(1) $$\{(x,y)| \quad x \in X \quad \& \quad y \in Fx\}.$$

Each ordered pair (x,y) in (1) is an ARC of the graph; x is the INITIAL POINT of the arc and y the TERMINAL POINT. An arc (x,x) whose initial point and terminal point coincide is a LOOP. (There are no loops in the functor-argument graphs in this book and each graph has a finite number of vertices.)

If $A \subset X$ and F_A is the restriction of F to A then a SUBGRAPH (A,F_A) of the graph (X,F) is the set:

(2) $$\{(x,y)| \quad x \in A \quad \& \quad y \in A \cap Fx\}.$$

The INDEGREE of a vertex x is the number of arcs having x as a terminal point (the number of arrows with tips at x in the diagram). The OUTDEGREE of a vertex x is the number of arcs having x for an initial point.

A PATH is a sequence of arcs such that the terminal point of each arc coincides with the initial point of the following arc.

[1] The terminology and definitions in this section are essentially the same as in Berge, *The Theory of Graphs and Its Applications.*

If there is a path from a vertex x to a vertex z then z is said to be REACHABLE from x. A CIRCUIT is a finite path such that the initial point of the first arc coincides with the terminal point of the last arc.

Two vertices x and y are said to form an EDGE [x,y] in a graph if (x,y) or (y,x) is an arc of the graph. Thus in (3) there are three arcs and two edges:

(3)

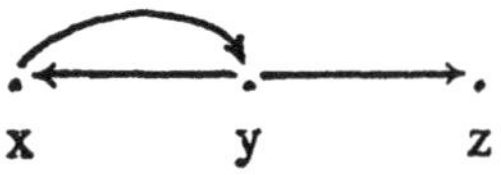

A sequence of edges, say $e_1, e_2, \ldots, e_n$, such that e_i shares one vertex with e_{i-1} and the other vertex with e_{i+1} is called a CHAIN. A CYCLE is a finite chain beginning and ending at the same vertex. (4) is a cycle, but not a circuit:

(4)

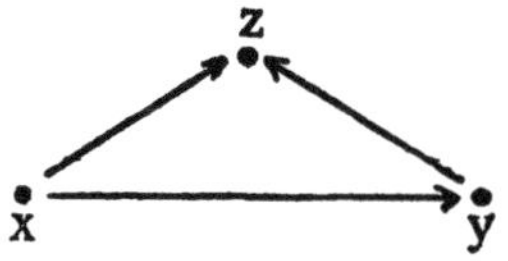

A subgraph of (X,F) is a COMPONENT of (X,F) if:

(i) there is a chain connecting any two distinct vertices in the subgraph (i.e., the subgraph is *connected*);
(ii) there is no chain connecting any vertex in the subgraph with a vertex not in the subgraph.

If a finite graph with at least two vertices has only one component and no cycles, it is called a TREE. (Of particular interest here is a tree in which there is a unique vertex (the ROOT) from which every other vertex is reachable, the indegree of the root being 0 and the indegree of every other vertex 1. If the structure assigned to a sentence yields this type of functor-argument graph there is no argument sharing involved.)

A vertex is said to be PENDANT if there is only one edge [x,y] containing it.

The definition of a graph given in (1) does not permit two distinct arcs with the same initial and terminal points as in (5):

(5)

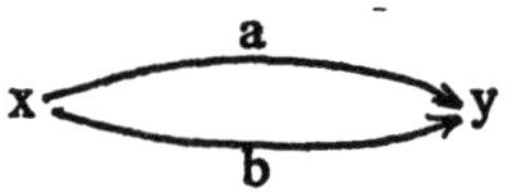

Let us call a and b PARALLEL ARCS. If our definition of a graph is extended to permit as many as s parallel arcs we may refer to the result as an s-graph.

BIBLIOGRAPHY

Ajdukiewicz, Kazimierz

1935 "Die syntaktische Konnexität", *Studia Philosophica* 1, 1-27. English translation in *Polish Logic 1920-1939* (Oxford University Press, 1967), ed. by Storrs McCall.

Bar Hillel, Yehoshua

1950 "On syntactical categories", *The Journal of Symbolic Logic* 15, 1-16.

1953 "A quasi arithmetical notation for syntactic description", *Language* 29, 47-58.

1961 "Some linguistic obstacles to machine translation", *Proceedings of the second international congress on cybernetics* (Namur), 197-207.

1960 with C. Gaifman and E. Shamir, "On categorial and Phrase Structure Grammars", *The Bulletin of the Research Council of Israel* 9F, 1-16. (The above articles are reprinted in *Language and Information* [Addison-Wesley, 1964], selected essays by Y. Bar-Hillel.)

Berge, Claude

1958 *The Theory of Graphs and its Applications* (Paris: Dunod), English translation (London: Methuen & Co., 1962).

Bocheński, I. M., O.P.

1949 "On the Syntactical Categories", *The New Schoolmen*, 259-280.

Carnap, Rudolf

1937 "Testability and Meaning", *Philosophy of Science* 4, 1-40.

Chomsky, Noam

1957 *Syntactic Structures* (= *Janua Linguarum*, series minor 4) (The Hague: Mouton).

1961 "On the notion 'rule of grammar'", *Structure of Language and Its Mathematical Aspects*, ed. by R. Jakobson (Providence, R. I.: American Mathematical Society), 6-24.

1965 *Aspects of the Theory of Syntax* (Cambridge, Massachusetts: M.I.T. Press).

Chomsky, N. and G. A. Miller

1963 "Introduction to the Formal Analysis of Natural Languages", *Handbook of Mathematical Psychology*, vol. II (New York: Wiley), 269-322.

Cohen, Joel M.

1967 "The Equivalence of Two Concepts of Categorial Grammar", *Information and Control* 10, 475-484.

Curry, Haskell B.
1961 "Some logical aspects of grammatical structure", *Structure of Language and Its Mathematical Aspects*, ed. by R. Jakobson (Providence, R.I.: American Mathematical Society), 56-68.

Goodman, Nelson
1961 "Graphs for linguistics", *Structure of Language and Its Mathematical Aspects*, ed. by R. Jakobson (Providence, R.I.: American Mathematical Society), 51-55.

Harman, Gilbert
1963 "Generative grammars without transformational rules: a defense of phrase structure", *Language* 39, 597-616.

Harris, Zellig S.
1951 *Structural Linguistics* (University of Chicago Press).
1957 "Cooccurrence and transformation in linguistic structure", *Language* 33, 293-340.
1962 *String Analysis of Sentence Structure* (The Hague: Mouton).
1963 *Discourse Analysis Reprints* (The Hague: Mouton).
1965 "Transformational theory", *Language* 41, 363-401.
1968 *Mathematical Structures of Language* (New York: Interscience, Wiley).
1970 *Papers in Structural and Transformational Linguistics* (Dordrecht: D. Reidel).
1953 with C. F. Voegelin, "Eliciting in linguistics", *Southwestern Journal of Anthropology* 9, No. 1, 59-72.

Hiż, Henry
1960 "The Intuitions of Grammatical Categories", *Methodus*, 311-319.
1961a "Congrammaticality, batteries of transformations and grammatical categories", *Structure of Language and Its Mathematical Aspects*, ed. by R. Jakobson (Providence, R.I.: American Mathematical Society), 43-50.
1961b "Syntactic Completion Analysis", *Transformations and Discourse Analysis Papers* 21 (University of Pennsylvania).
1964 "A Linearization of Chemical Graphs", *Journal of Chemical Documentation* 4, 173-180.
1967 "Grammar Logicism", *The Monist* 51, 110-127.
1968 "Computable and Uncomputable Elements of Syntax", *Logic, Methodology and Philosophy of Sciences III*, ed. by Rootselaar and Staal (Amsterdam: North-Holland), 239-254.
1969 "Referentials", *Semiotica*, 136-166.
1971 "On the Abstractness of Individuals" *Identity and Individuation* (New York University Press), 251-261.

Hockett, Charles
1958 *A Course in Modern Linguistics* (New York: Macmillan).
1968 *The State of the Art* (The Hague: Mouton).

Husserl, Edmund
1913 *Logische Untersuchungen*, 2nd edition (Halle).

Jespersen, Otto
1924 *The Philosophy of Grammar* (London: Allen & Unwin).

Lambek, Joachim

1958 "The mathematics of sentence structure", *American Mathematical Monthly* 65, 154-170.

1961 "On the calculus of syntactic types", *Structure of Language and Its Mathematical Aspects*, ed. by R. Jakobson (Providence, R.I.: American Mathematical Society), 166-178.

Leśniewski, Stanisław

1929 "Grundzüge eines neuen Systems der Grundlagen der Mathematik", *Fundamenta Mathematicae* 14, 1-81.

Lieberman, Philip

1967 *Intonation, Perception, and Language* (Cambridge, Massachusetts: M.I.T. Press).

Robbins, Beverly

1968 *The Definite Article in English Transformations* (The Hague: Mouton).

Smaby, Richard

1971 *Paraphrase Grammars* (Dordrecht Holland: D. Reidel).

Tarski, Alfred

1956 "The Concept of Truth in Formalized Languages", *Logic, Semantics, Metamathematics*, translated by J. H. Woodger (London: Oxford University Press), 152-278.

Trager, George and H. L. Smith

1951 *An Outline of English Structure* (= *Studies in Linguistics*, Occasional Papers 3) (Norman, Oklahoma: Battenburg Press).

INDEX

Bei Fragen zur Produktsicherheit wenden Sie sich bitte an:
If you have any questions regarding product safety,
please contact:

Walter de Gruyter GmbH
Genthiner Straße 13
10785 Berlin
productsafety@degruyterbrill.com